Walking the Wetlands

Wiley Nature Editions

At the Water's Edge: Nature Study in Lakes, Streams and Ponds, by Alan M. Cvancara

Mountains: A Natural History and Hiking Guide, by Margaret Fuller

The Oceans: A Book of Questions and Answers, by Don Groves

Walking the Wetlands: A Hiker's Guide to Common Plants and Animals of Marshes, Bogs and Swamps, by Janet Lyons and Sandra Jordan

WALKING THE WETLANDS

A Hiker's Guide
to Common Plants and Animals
of Marshes, Bogs, and Swamps

Janet Lyons
and Sandra Jordan

Illustrated by Ron Schneider

Wiley Nature Editions

John Wiley & Sons, Inc.

New York • Chichester • Brisbane • Toronto • Singapore

Publisher: Stephen Kippur
Editor: David Sobel
Managing Editor: Frank Grazioli
Compositor: Crane Typesetting Service, Inc.

Library of Congress Cataloging-in-Publication Data

Lyons, Janet.
Walking the wetlands : a hiker's guide to common plants and animals of marshes, bogs, and swamps / Janet Lyons and Sandra Jordan.
p. cm.
Bibliography: p.
ISBN 0-471-62087-4
1. Wetland fauna—United States—Identification. 2. Wetland flora—United States—Identification. I. Jordan, Sandra, 1947– II. Title.
QH104.L96 1988b
574.973—dc19 88-36672
CIP

Printed in the United States of America

89 90 10 9 8 7 6 5 4 3 2 1

To Dr. Lee Emrich, Dr. Allen Marchisin, and Dr. Anthony Palumbo, for their technical review of the manuscript; to Dr. C. Thomas Lyons, for his continuing encouragement; and to Kerry, Kelly, and Kasey, for their enthusiasm and support.

Contents

List of Illustrations

Foreword

For too long wetlands have been the forgotten stepchild of our environment.

Joyce Kilmer never wrote, "I think that I shall never see, a poem as lovely as a swamp." Joni Mitchell didn't sob, "I've looked at bogs from both sides now." Although my children tell me that Bruce Springsteen has touched a few hearts by singing about the "swamps of Jersey," most artists, as well as naturalists and public officials, either ignored wetlands or deemed them useless and replaceable.

Now we know better. Now we are beginning to realize that swamps and bogs are not only troves of infinite beauty and variety, but that they are the fragile linchpins of our ecosystem. Without wetlands, many birds, plants, and animals would simply vanish from the planet.

Sandra Jordan and Janet Lyons have captured the splendor and importance of freshwater wetlands in this book. In a breezy, informal style (for example, the authors call the mosquito, "the vampire of the marsh"), Janet and Sandra pull back the veil that has obscured these natural gold mines. You will learn why the Iris flower is able to cross pollinate; how a dragonfly can fly sideways at sixty miles an hour; and how a barred owl will occasionally let loose with a blood-curdling scream to send chills down your spine.

The next time you have an hour or two on your hands, take a walk through a nearby swamp, marsh, or bog. Bring this book along. I guarantee that you will learn a hundred new things about the world around you. You'll end the walk with a new appreciation for the wonders of mother nature.

But don't let that good feeling end there. In many parts of America, wetlands are disappearing daily beneath the bulldozer. Hundreds of acres of swamps and bogs are being turned into condominium centers and office complexes. As anyone who reads this book will appreciate, that is a senseless and shortsighted tragedy.

So when you have finished *Walking the Wetlands,* pick up a pen of your own. Write your Congressman or Governor and let them know that you understand that swamps and bogs are more than mucky water and annoying insects. Tell them about the long-necked Anhinga and the Jack-in-the-Pulpit. Tell them to make it a priority to protect wetlands from uncontrolled development. If they ask why, tell them to read this book.

Thomas H. Kean, Governor of New Jersey,
Chairman, National Panel
on Wetlands Policy.

Preface

Intimacies of the Swamp

When the burden of being stretches its bounds
Escape to the swamp toward ethereal sounds.
Let go your senses to its soggy spirit.
A mystical magic transcends those near it.
Feel the silent, invisible movement
of elves on secret missions sent,
Passing ghostly sentinels from warmer days
Glistening now in a winter haze.
Behold the orchestrated melodies rare
Blending fragrances in the dampened air.
Go forth renewed, the passions spent.
The wetlands legacy—a soul content.

Introduction

Today the naturalist, professional and amateur alike, recognizes the wetlands as an unrivaled site for partaking of nature's choicest offerings. There, it is indeed likely that you will experience a sense of reverence associated with the sights and sounds of nature. From the rhythmic drumming of a pileated woodpecker to the gentle flutter of a blue azure butterfly, our senses are aroused.

The very nature of a water-laden environment guarantees that an abundance and enormous diversity of life forms will be found there. Our freshwater wetlands provide a footing for thousands of resident species, both plant and animal; provide water for upland species; provide cover and moisture for nest sites and subsequent young; and provide vital rest areas for migratory species. The variety of organisms found exclusively in such a habitat defies enumeration.

Come along as we walk the wetlands.

The purpose of this book is to guide the reader in identifying some of the most abundant and some of the most unusual organisms of freshwater wetlands of the United States. The term *wetlands* refers to marshes, swamps, and bogs that may be holding fresh, salty, or brackish water. This book concerns itself exclusively with the freshwater wetlands. Each organism is profiled by a pen-and-ink sketch and a description of its characteristics, range, and habitat. Identification in this manner becomes a delight rather than a chore. A detailed, informative account then follows to enhance an awareness and appreciation of the species.

Bogs, swamps, and marshes cover seventy million acres of the United States. Sculpted by relentless geological processes, these waterlogged depressions may vary in size from small, temporary pools to vast tracts covering thousands of acres. These are dynamic areas wrought with cyclical flooding. The plants that flourish are those which can tolerate frequent extreme changes in the moisture content of the soil. The animals found here are innumerable, and as with the plants, diversity abounds.

Wetlands are unique areas that have a personality of their own, one that changes greatly from season to season. Wetlands may be virgin areas, or they may contain man-made boardwalks. At first the diversity of life may seem overwhelming, but visitors who come to know these areas soon find that the extent of participation determines familiarity. In order to enjoy its bounty, however, one must be prepared.

When clouds of breath signal a cold, crisp day, be cautious of easily chilled fingers and toes. Winter conditions demand a warm jacket and pants, waterproof boots, a hat, and gloves. A flashlight and compass may prove invaluable if one finds sunset approaching more rapidly than expected. Binoculars and a hand lens will magnify your pleasure.

It is during this time that the swamp appears lifeless. The subdued sunlight reflects off the glistening branches, straining under the weight of ice and snow. Most creatures are resting, snugly encased beneath the thick, white blanket or hidden in crevices. Some organisms, however, like the birds, are in constant motion. In spite of the cold, wintering songbirds add a delightful melody and a splash of color to the bleak, gray landscape. Birds of prey, normally obscured by the dense vegetation, are strikingly evident. Even owl activity is apparent on an early moonlit evening. Prepare for lengthy periods of inactivity spent in quiet observation. And bear in mind that frigid conditions may prevail.

As winter loosens its grip, the familiar frozen pathways may be short-circuited by melting ice and spring rains. Now high, warm, waterproof boots are a must. Children are especially vulnerable to icy water slipping over their boot tops. During this season a warm, waterproof jacket, a hat, and gloves may still be necessary, depending upon the latitude. Accessories for viewing, photographing, or identifying specimens or just enjoying time are limited only by individual preference. It is now that the swamp begins to hum with life. Peeping frogs, a few buzzing insect swarms, and uncurling fiddleheads affirm the end of winter. The air smells sweet and earthy as each wildflower lends its unique fragrance. Water is everywhere, and you may find your travels encumbered by deep water and, in many cases, mud.

As summer approaches and the wetlands' color changes from a bright yellow-green to the deeper green of summer, the pools begin to recede. Now the warm, humid air is alive with insects, many of them blood-seeking. Plenty of insect repellent, long sleeves, and a hat will do wonders for your tolerance. The wetlands now are densely vegetated, making movement more difficult. Boots may give way to sneakers for more mobility, but mud, if not standing water, will still be present. The trees are filled with birds, flowers are everywhere, and mammal, amphibian, and reptile tracks abound. While the temperature may tempt one to wear shorts, exposed legs are at the mercy of such hazards as preying insects, poison ivy, and dried undergrowth, to mention just a few.

The air grows crisp and the humidity decreases with the onset of autumn. As the area empties of many songbirds, the poison ivy, creeper, and sumac begin to show a brilliant crimson, signaling to the hungry travelers of the skies. Now migratory waterfowl are frequent visitors stopping to rest and eat. The insect population is decreasing rapidly, but on warm days it revives almost to capacity.

This is the season of the hunter, so dress in bright colors. Be sure that the area you are exploring is posted and that someone knows of your whereabouts. A light jacket and hiking boots make a practical outfit.

As one begins to explore the wetlands, their tremendous variation becomes apparent. Precise categorizing of swamps, bogs, and marshes can be rather elusive, for they are generally separated by vague limits, dependent mostly upon the characteristic flora and fauna of the area.

Bogs are born of steep-sided depressions of glacial origin. They characteristically have poor or virtually no drainage. The cushiony mats of vegetation are primarily sphagnum and hypnum mosses. As these and other plants partially decay, thick beds of peat are formed. The acids from the accumulating peat lend a brownish tinge to the water. This acidic condition coupled with a very low oxygen content (the result of minimal aeration) creates an alien environment to aquatic microflora. The few bacteria that can tolerate this high acid/low oxygen state have difficulty decomposing organic material. There are perhaps one hundred documented cases of human bodies hundreds of years old being removed from bog bottoms in almost perfectly preserved condition. The two-thousand-year-old Tollund man discovered in a Danish bog in 1950 is one such example.

The significance of bog water chemistry extends to more sophisticated flora as well. Minimal silting and reduced bacterial action limit the level of nutrients in the water. Bogs, therefore, are populated by a rather limited variety of vegetation and consequently fauna.

Low shrubs like leatherleaf and wild cranberry often grow into the dense hummocks of moss and sedge. The matting may become sturdy enough to support a man's weight. Jumping on such a massive growth can result in vegetative movement many yards away. This is the basis of the quaking bog.

The trees most likely to encroach upon the margins of a bog and eventually toward its center are the larch and spruce. Unlike most upland trees, these can tolerate both the water chemistry and the peat-based soil.

Marshes originate from the slow filling in of ponds and lakes. Tall grasses along the edge build soil and creep inward. The evolving "aquatic meadow" is usually covered with water varying in depth from several inches to several feet. Often the drainage is quite poor, but its very existence accounts for the all-important aeration. Any water turbulence increases the dissolved oxygen, making available to microorganisms a vital element of their metabolism. These metabolic processes produce vast quantities of nutrients that are released into the water. Those nutrients are then available to both free-living and rooted life forms.

The most common plants are representatives of the reed, grass, cane, and rush families. Animal groups may include small mammals, waterfowl, reptiles, amphibians, and a variety of invertebrates.

Swamps have greatly fluctuating water levels which are controlled by seasonal weather patterns. As silting and the decaying organic material build soil, larger and more varied plants begin to gain footing. Dominant plants include a wide range of wildflowers and ferns. Water-tolerant trees such as dogwoods, alders, and willows begin to root along the marsh edges. As this progression occurs the wooded swamp evolves. This developing habitat attracts a diversity of aquatic animals, as well as highland species feeding along the margins. Thus, the swamp represents a natural succession from the marsh to the forest.

Politically and culturally the "uselessness" of our wetlands is finally being questioned. A valuable wildlife habitat, the wetlands are the exclusive home of many reptiles and amphibians, including such endangered species as the bog turtle and the eastern salamander. For innumerable waterfowl, there is no better place for feeding, and the cover provided is essential to successful nesting.

It is already well documented that habitat loss, especially breeding grounds, through commercial filling of swamps and marshes is responsible for the rapid decline of many species. Beyond satisfying the needs of their own residents, wetland tracts also provide an Eden for migratory animals. This consideration significantly expands the importance of such habitats.

During periods of excessive rain, storage basins retain water for slow release, thus vastly reducing erosion. The slow seepage into aquifers and watersheds allows time for the vegetation to entrap pollutants rich in nitrogen and phosphorus. Sediments are filtered and valuable nutrients absorbed. These nutrients ultimately resurface as a rich algal bloom to be consumed by aquatic organisms. Flood control and water purification, therefore, are essential contributions of our freshwater wetlands.

The recreational significance of the wetlands is immediate if one considers nature photography or simply nature appreciation, birding, and hiking. Extended indirectly as they affect total watersheds, these areas also figure significantly in water sports, fishing, hunting, and trapping. Most importantly, perhaps, they represent one of the earth's enchanted natural areas, the essence of which cannot be replaced. To belittle these tracts as wastelands devalues the human heritage.

Walking the Wetlands

British Soldier and Pixie Cup Cladonia

Cladonia sp.

DESCRIPTION

Classification: Lichen
Size: Height to 1″
Characteristics: Pale blue-green; *cristatella*—red, knobby fruiting body; *pyxidata*—brown, cup-shaped fruiting body

HABITAT

Swamps, marshes, bogs

RANGE

Continental U.S.A.

PERCHED demurely upon a rotting log or equally at ease on bare stone, the pale gray-green *Cladonia* has long been recognized as a "pioneer plant." Lichens represent the first vegetative form to invade a barren frontier. These organisms will attach to rock surfaces and begin the timeless process of soil formation. Even in the wetlands, an extensively interacting living system, this contribution is significant.

The lichen body is a mutually beneficial union of a fungus and an alga. This self-supporting relationship is considered by many to be a perfect example of symbiosis. The algal cells photosynthesize and share their product with their fungal associates. The fungus, able to absorb ten to thirty times its weight in water, maintains an adequate water supply for the entire plant. Interestingly, some seventeen thousand species of fungi have established similar relationships, while the same is true of only twenty-five species of algae.

Cladonia cristatella, commonly called British soldier or scarlet-crested lichen, holds knobby red fruiting tips on one-inch pale green stalks. The colors are most striking in the early spring. The stalks of *Cladonia pyxidata* are similar in size but are scaly and trumpet-shaped. Brown, disk-shaped fruiting bodies grow along the lip of the cup.

Although lichens grow more slowly than any other plants, they serve as a major food source for a variety of organisms. In the northern reaches of the earth, where food is often scarce, lichens are eaten by animals and people alike. In the wetlands habitats snails, slugs, and many insects regularly feast upon these spongy plants.

Chemists have long used litmus, an extract from lichens, to indicate the acidic or alkaline nature of a substance. During the Middle Ages, lichens were classified as an effective medicinal herb. From the ancient Greek and Roman civilizations to present-day European cultures, lichens have been collected in vast quantities for fabric dyes. Dyes processed from the lichens are used in the manufacture of genuine Harris tweeds.

Thus these plants, among the simplest residents of our planet, have served us in a multitude of ways through the ages. Unfortunately, the hardy lichen is especially sensitive to air pollution and is beginning to disappear from many areas. It is sad to think that these venerable, versatile organisms may succumb to man's progress.

Three-Toothed Bazzania

Bazzania trilobata

DESCRIPTION

Classification: Liverwort
Size: Length to 5″; less than ½″ wide
Characteristics: Three-toothed leaves, densely packed and overlapping along the stem

HABITAT

Swamps, marshes

RANGE

Eastern half of the U.S.A.

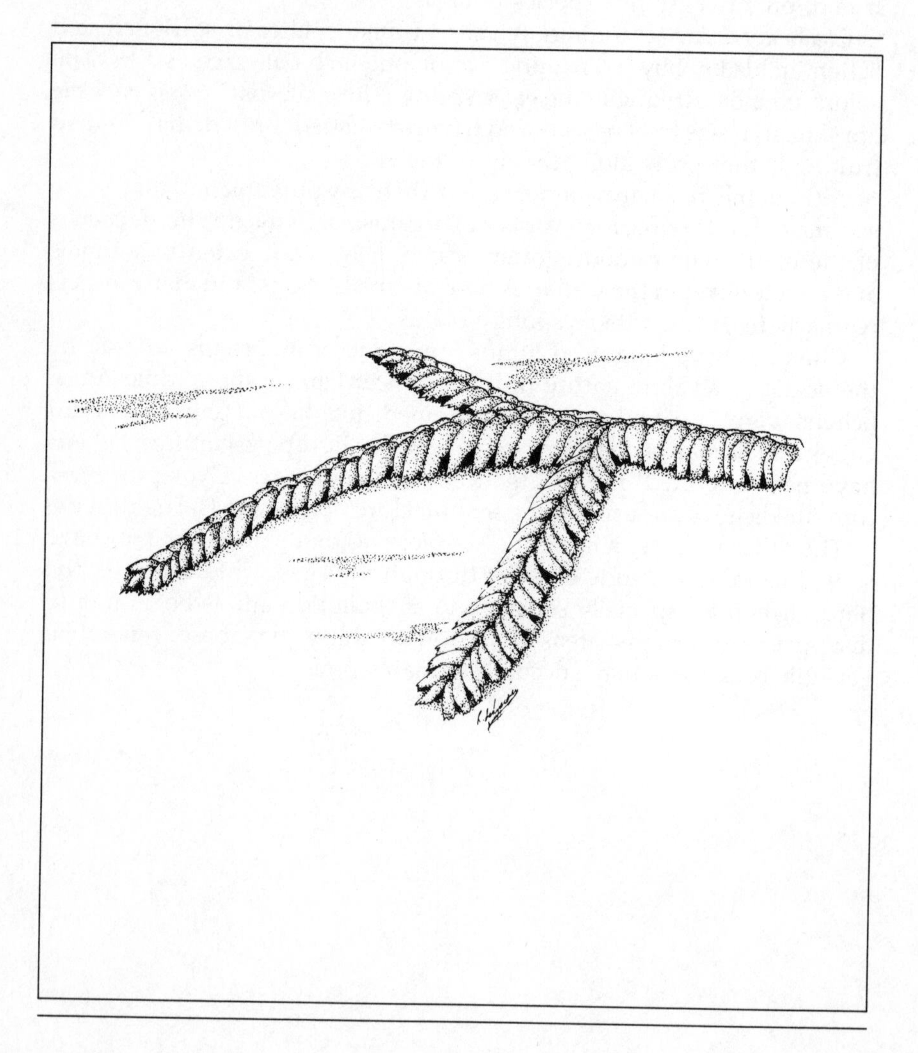

THE liverworts are a fascinating group of plants frequently eclipsed by the more ostentatious wetlands inhabitants. To our very imaginative ancestors, the leaves of the liverworts resembled the lobes of a human liver. In the early days of herbal medicine, it was reasoned that any plant resembling a body organ could cure an ailment of that organ. Today the liverworts retain their obsolete epithet although they are distinguished by no known economic value.

The three-toothed bazzania is one of the largest of the leafy liverworts. The flat, green, overlapping leaves are arranged in two rows which are visible from above. A third row of more coarsely toothed leaves runs along the length of the stem on the underside. While an individual plant rarely exceeds five inches in length, colonies of the prostrate, creeping bazzania commonly form mats of several square feet.

Typifying liverwort structure, the bazzania has no true stems, roots, or leaves. Water and carbon dioxide are absorbed through the pores or the upper cell layer. Conducting water with negligible efficiency, liverworts are restricted to an amphibious existence.

Tiny hairlike rhizoids issue from the underside of the plants. Acting as primitive roots, these fragile threads form an attachment to soil, rock crevices, or an emergent tree trunk.

Pointed Mnium

Mnium cuspidatum

DESCRIPTION

Classification: Moss
Size: Height to 1½″
Characteristics: Pale to dark green leaves with distinctive points on tips

HABITAT

Swamps, marshes

RANGE

Eastern half of the U.S.A.

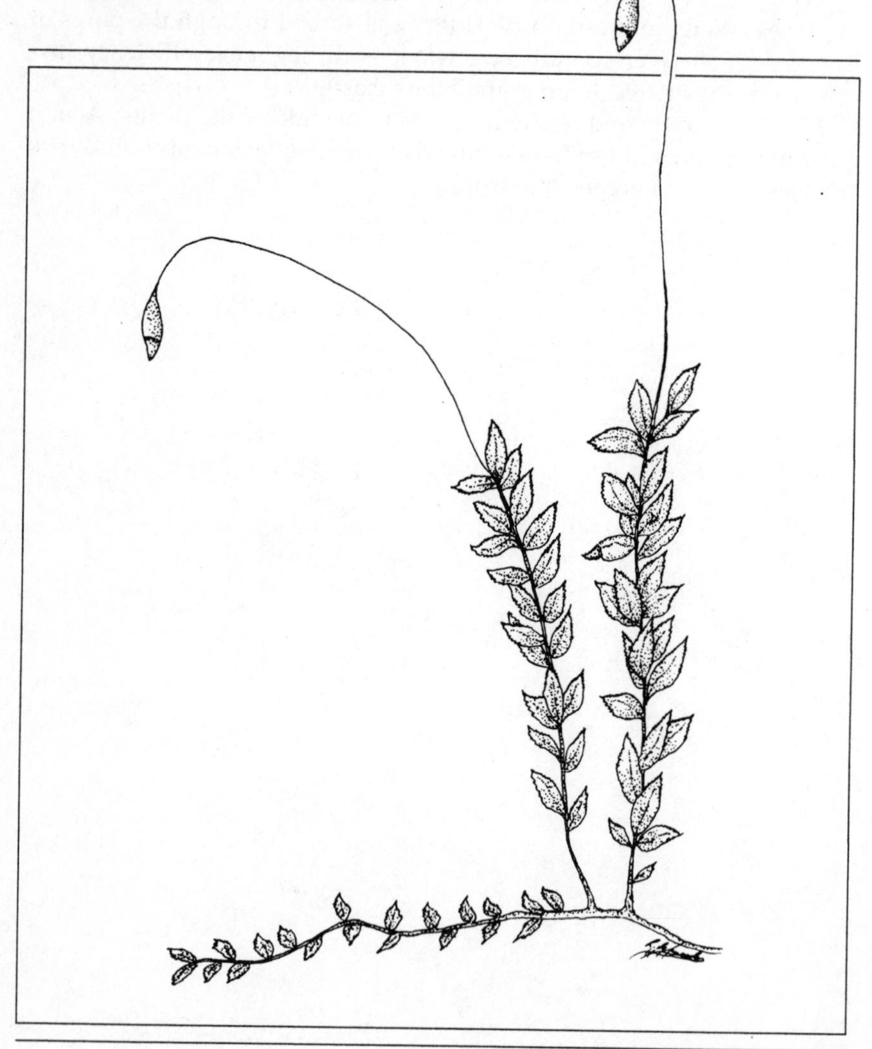

IN the overall grandeur of the wetlands, the unobtrusive pointed mnium may easily go unnoticed. Perched near the shaded water's edge in near-saturated soil, its pale leaves glisten in the splintered sunlight. This tiny plant is, in fact, the hub of a rather complex series of processes.

Careful scrutiny reveals rather attractive foliage. The tiny pointed leaves are unusually broad for a moss plant. Clustered about the central stalk and arched outward, the leaves become conspicuously larger as they approach the tip of the stem. In the male plants, this array culminates in a delicate rosette.

The reproductive cycle of the pointed mnium is by no means unique among the mosses. During the spring, a wiry stalk towers above the foliage, exposing a nodding green capsule filled with developing spores. Well before the sultry summer weather, the spore case matures, releasing the now ripened spores. Carried randomly by wind or water, spores landing on adequately moist soil will germinate into a green, threadlike mass. Tiny buds soon arise and leaf out into the foliage that we recognize as pointed mnium.

A single plant will produce both male and female structures. Sperm mature within the male rosette and literally swim through the moisture of spring rains, groundwater, or even the morning dew. Those sperm which fortuitously advance to the female plant will fertilize the awaiting eggs.

The fertilized eggs give rise to a long, wiry stem topped with a nodding case filled with spores. Thus each type of germ cell gives rise to plant parts quite unlike those that produced them.

Mosses, like the pointed mnium, are perennial. Each spring new growth emerges from the previous year's plant. Runners, which also give rise to new shoots, are continually being sent out from beneath the soil of these plants, further extending their range.

Inhabiting areas harassed by the erosive nature of water, pointed mnium provides an important environmental service in both building and binding soil.

Sphagnum Moss

Sphagnum capillaceum

DESCRIPTION

Classification: Moss
Size: Height to 1½″
Characteristics: Pale green branching stems; pointed leaves

HABITAT

Swamps, marshes, bogs

RANGE

Continental U.S.A.

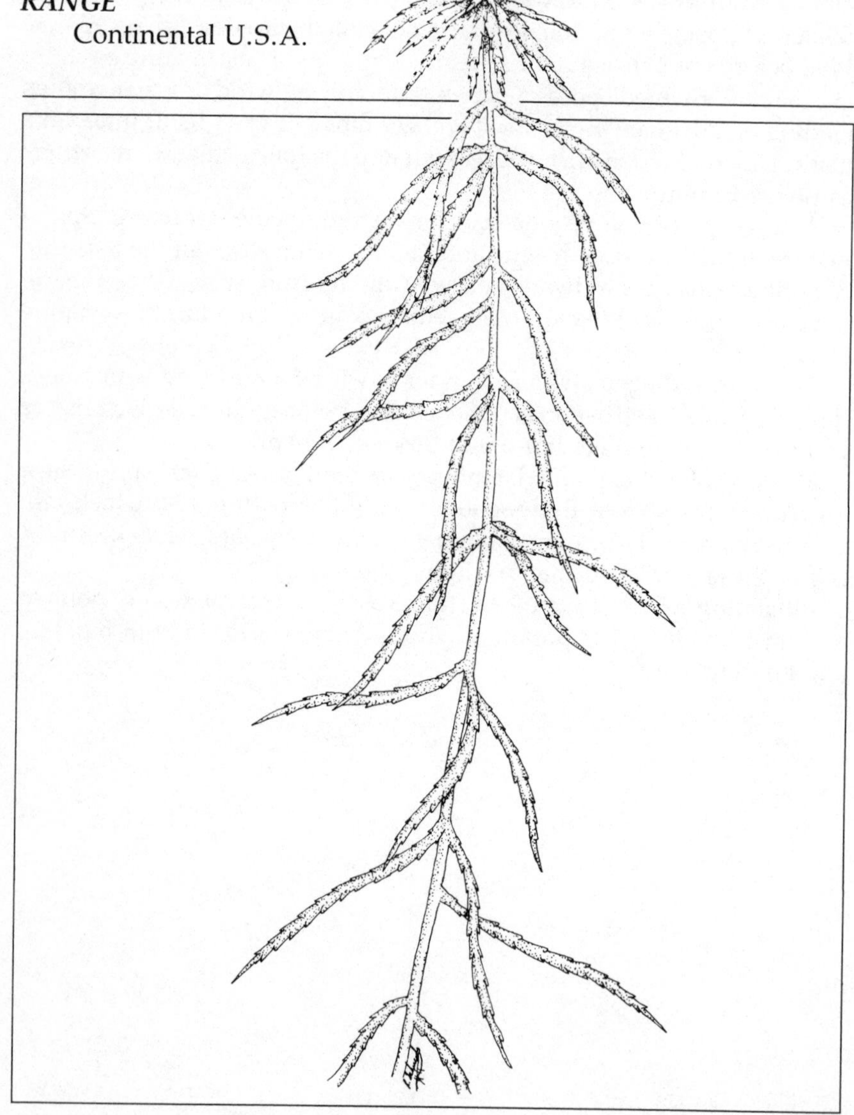

Of all the wetlands mosses, none has a greater impact than the abundant sphagnum moss. The spongy hummocks dapple the water's surface, providing a support for an alighting insect, a retreat for a roving frog, or a foothold for an optimistic fiddlehead. The plant structure is a vast network of small and large cells, both living and dead. It is this very structure which gives sphagnum moss its incredible capacity to hold water. Estimated at soaking up over one hundred times its own weight in water, sphagnum moss far exceeds the absorbent potential of cotton.

The general condition of the plant depends heavily upon the nature of the wetlands that it inhabits. In swamps and marshes which are rich in dissolved oxygen and bacterial activity, the lower portion of the hummock is in a state of constant decay, as new growth continues to emerge from above.

In bogs, where tannins and organic acids maintain a nearly sterile environment, the dying undergrowth does not decay. A raft of substantial thickness, strong enough to support a man's weight, forms the dense growth. The springy response of the sphagnum moss under normal walking pressure has given rise to the term *quaking bog*.

As the dead plant matter accumulates without decaying, the weight from above compresses the organic debris. In Great Britain and other parts of the world, this compressed sphagnum moss is cut into blocks, dried, and used as fuel. These peat blocks can form from living plants in as little as twenty-five years.

During World War I, sphagnum moss from bogs was used as a surgical dressing in place of cotton because of its natural sterility and absorbent qualities. Today sphagnum moss (also called peat) is used to keep packaged plants moist during shipping. Dried, shredded, and packed in bales, peat moss is essential to gardeners as an aid to retaining moisture in poor soil.

Cinnamon Fern

Osmunda cinnamomea

DESCRIPTION

Classification: Fern
Size: Height to 3′
Characteristics: Fertile fronds clublike, pointed, cinnamon-colored; sterile fronds taller, more erect, waxy green

HABITAT

Swamps

RANGE

Eastern half of the U.S.A.; Texas; New Mexico

DURING the chilly days of early spring, the tightly curled fiddleheads of the cinnamon fern proclaim the coming thaw. They are cozily enveloped in a thick, woolly coat of silvery hair. As the sun gently warms the earth, these coats are shed and the fiddleheads unfurl.

The first sparse stand of fertile fronds that are produced offer sharp contrast to the luxuriant growth that is to follow. Bright green at first, these fronds will turn cinnamon brown by early summer. Clinging like miniature grape clusters to the rigid stalks are the immature spore capsules.

The taller, waxy green sterile fronds arise as the spores mature. These large, strong, twice-cut fronds grow in a circular cluster, forming an arching stronghold around the fertile fronds. Reddish brown tufts of hair found at the base of each leaflet give the fern its name. Prized by the ruby-throated hummingbird, these cinnamon tufts are used to line its nest.

The extravagant bouquetlike growth of the cinnamon fern is magnificent. However, when the fern is exposed to such stresses as extreme weather conditions or fire, its fronds become stunted and stiff and lose their arching curve.

Those interested in edible plants will find that the young fiddleheads can be a tasty addition to a meal. Stripped of their woolly coats, they can be used in a salad or as a cooked vegetable.

Marsh Fern

Thelypteris palustris

DESCRIPTION

Classification: Fern
Size: Height to 4′
Characteristics: Delicate, narrow-leaved fern with exceptionally long stalks; bright green

HABITAT

Swamps, marshes, bogs

RANGE

Maine to Florida; west to Oklahoma

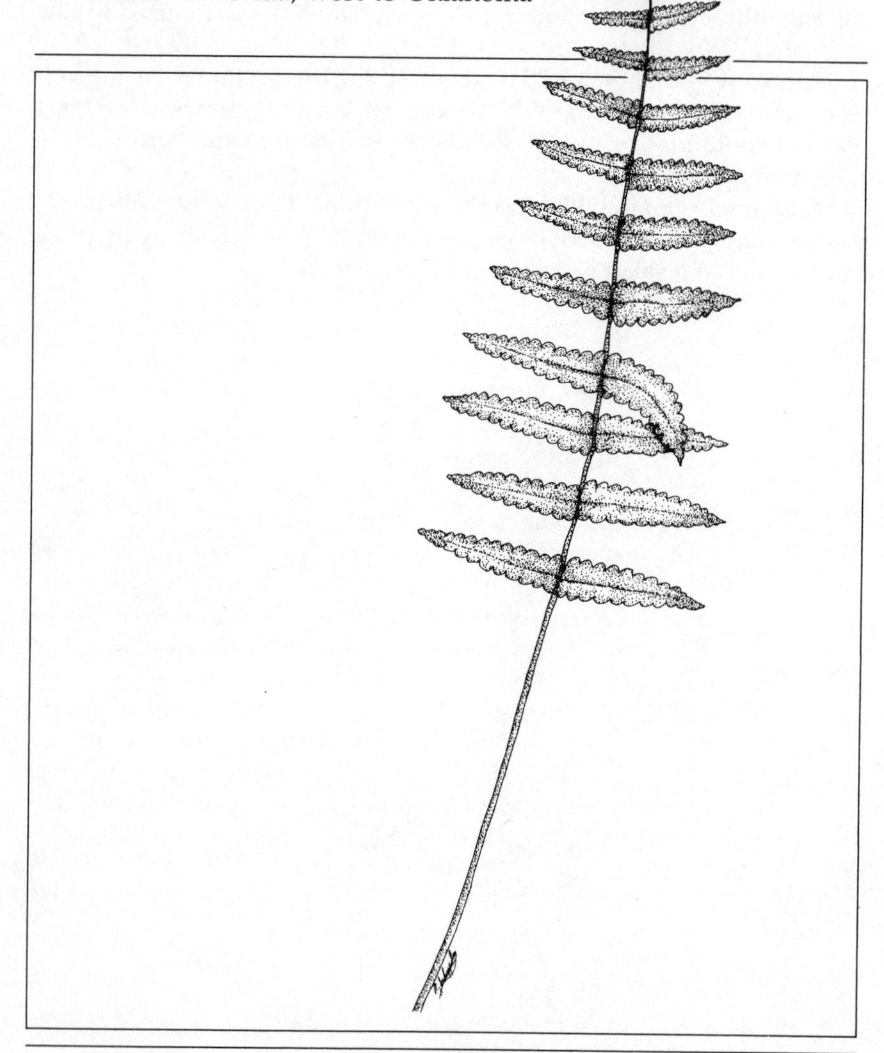

GRACING the wetlands, the delicate, bright green marsh fern displays its finery among the many colorful marsh flowers. From the brilliant gold of the marsh marigold to the subtle hues of the blue flag, the diverse shades are enhanced by the vivid green of the marsh fern. Commonly found in wet meadows, where it prefers rich muddy soil, it is often called the meadow fern.

The sensitivity of the marsh fern to its companions influences its growth. If grown in an area of low, dense vegetation, the fern will remain relatively short. On the other hand, medium-tall neighbors competing for sunlight promote additional growth in the fern. Under these conditions it may reach a height of four feet.

The twice-cut frond is divided into pinnae, which are further divided into deep, bluntly cut lobes. The leaves of the fern have exceptionally long stalks, even longer than the blades, as though the fern was determined to keep its dainty foliage dry.

The marsh fern is a plant whose leaves exhibit dimorphism. In the early spring, the fiddleheads unfurl to form masses of light green sterile leaves. By early summer these leaves have matured and darkened in color. Now newly formed yellow-green fertile leaves begin to emerge. The fertile leaves are more erect and soon surpass the sterile leaves. Because the tips of their fronds curl to form pointy ends, the fertile leaves impart a wilted look to the plant. This appearance is in direct contrast to the smooth, sturdy sterile leaf edges.

By midsummer the fertile leaves produce microscopic spores contained in tiny capsules called sporangia. The sporangia are clustered within a pale, kidney-shaped membrane called an indusium. Located on the back of the frond, the indusia are an important aid to fern identification.

By late summer the sporangia burst open, releasing thick clouds of spores into the air. Each spore that finds a suitable environment will grow into a minute, heart-shaped plant called a prothallus. Specialized cells on the prothallus eventually will give rise to swarms of sperm and eggs. As the sperm and egg unite, a new leafy fern is produced.

There are many legends surrounding the ferns, a favorite one being the legend of the fern's shiny magical seeds. These seeds, which could render a person invisible, were dispersed from tiny blue flowers only on Midsummer's Eve. In order to collect the seeds, groups of people journeyed into the mysterious marsh in search of a mystical encounter. This activity, known as "watching the fern," became extremely widespread. Such expeditions, considered too closely related to black magic, were banned by the Church.

Royal Fern

Osmunda regalis

DESCRIPTION

Classification: Fern

Size: Height to 6′

Characteristics: Pale green fronds resembling locust tree leaves; fertile leaves bear cinnamon brown sporangia clusters at tips

HABITAT

Swamps, marshes, bogs

RANGE

Maine to Florida; west to Texas

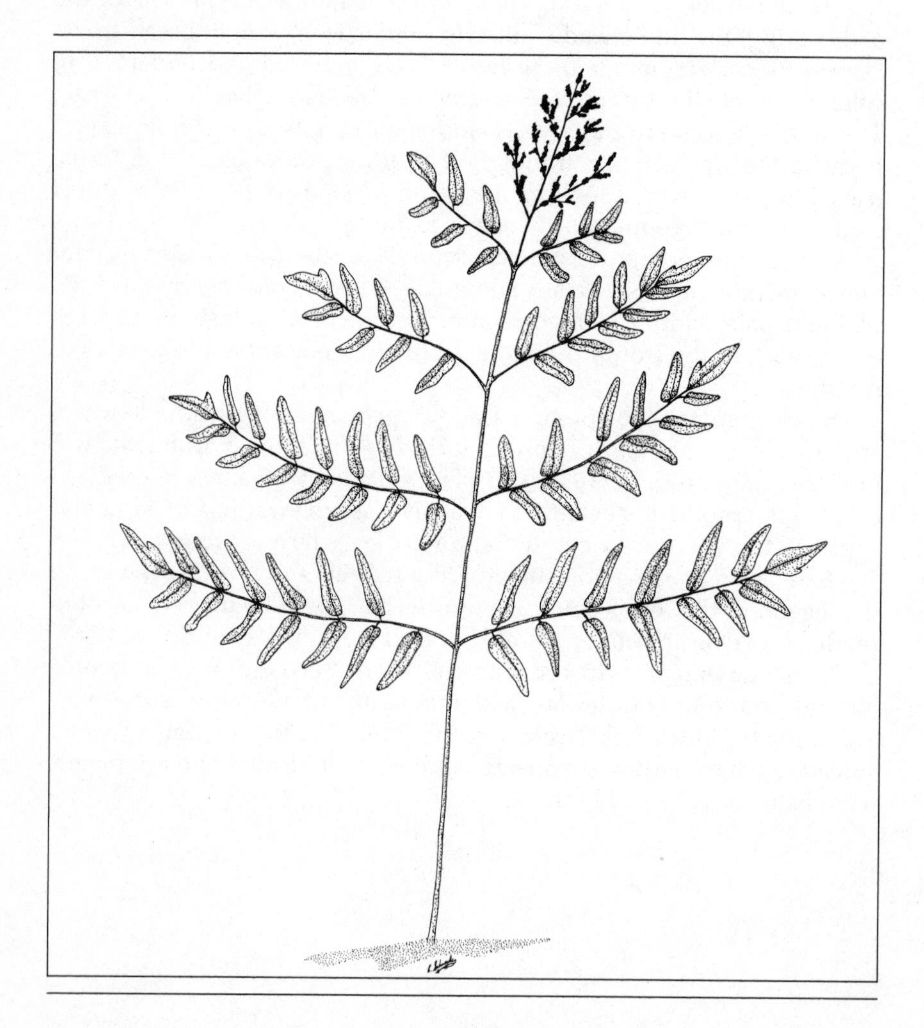

THE lush growth of the royal fern lends a junglelike quality to the wetlands. This fern was originally called the water fern because of its majestic growth in shallow water. The large, arching fronds arise from fragile, reddish brown fiddleheads in the early spring and unfurl daintily amid the vivid hues of the marsh marigold. Spreading slowly from one central plant outward in an enlarging circle, this twice-cut plant may reach a height of nearly six feet.

The large, light green sterile leaves of the royal fern form an elegant crown encompassing the fertile fronds. At a distance the plant is fernlike in appearance, but closer examination reveals pinnae resembling locust tree leaves. The fertile fronds have cinnamon-brown sporangia at their tips, and often this structure is mistaken for a flower. Thus another common name for the royal fern is the flowering fern.

Legend says that Osmunder, a Saxon god, hid his wife and child among the ferns to protect them from the marauding Danes. In later years, the fern was named *Osmunda regalis* in honor of this tale.

Mucilage from the stem of the royal fern was used in colonial times for treating coughs, but this practice is now obsolete. Care should be taken to preserve this stately plant for future generations.

Sensitive Fern

Onoclea sensibilis

DESCRIPTION:

Classification: Fern
Size: Height to 3′
Characteristics: Sterile fronds yellow-green, coarse, triangular-shaped; fertile fronds green, beadlike

HABITAT

Swamps, marshes

RANGE

Maine to Florida; west to Texas

PREFERRING swamps and wet meadows, this common deciduous fern grows in such rich profusion that it has often been called the weed of the fern family. In early spring, shallow rootstocks send up prominent masses of pale red fiddleheads. In an exuberant display of yellow-green foliage, the sterile fronds emerge proudly from the croziers.

This once-cut fern is rather coarse and heavily veined. The upper portion of the sterile frond has an uncut look. In contrast, the lower portion of each blade is cut deeply to the mid vein.

Lacking any leaflike structures, the fertile fronds resemble wands with rigid rows of green beads. These fertile fronds are produced in late summer. Each bead is a modified pinnule that has formed around several sori and will turn brown upon maturation. This beadlike spore-bearing section of the fertile frond has given the plant its alternate name of bead fern.

Amid the prostrate and dying sterile fronds, the brown-beaded stalks remain standing. Desiccated and decrepit, many will withstand the harsh winter snows. Spring will find the withered skeletons sharply contrasting with the fresh new growth.

In areas where this plant suffers from the assault of man or nature, it appears in an altered form known as the *Forma obtusilobata*.

Common Horsetail and Scouring Rush

Equisetum sp.

DESCRIPTION

Classification: Horsetail
Size: Height to 3′
Characteristics: Jointed, hollow, pale green stems with longitudinal grooves

HABITAT

Swamps, marshes

RANGE

Continental U.S.A.

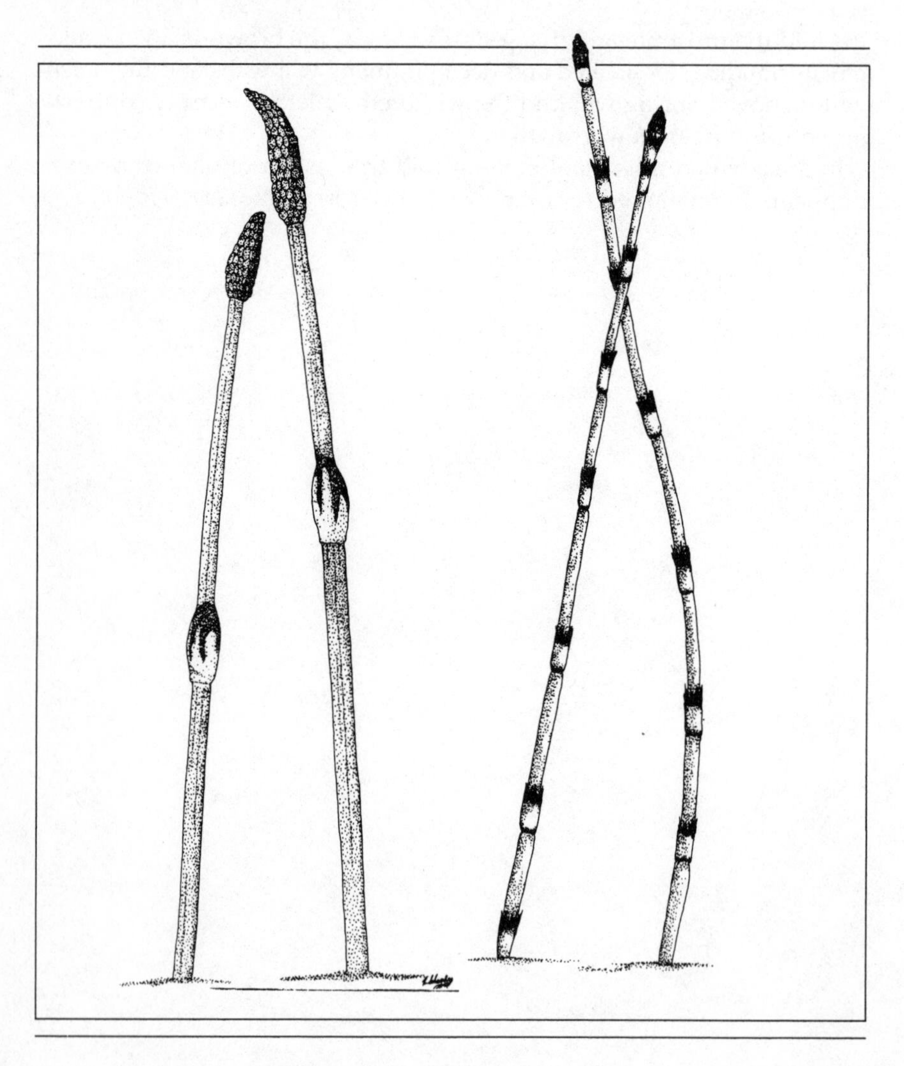

THE common horsetail and scouring rush can trace their lineage back 400 million years to the Devonian Period. Their now extinct, distant cousins were prominent plants of the Carboniferous swamps. Today's horsetails, bearing neither lush foliage nor colorful flowers, are not considered a significant part of the earth's vegetation.

Binding soil along water margins, the perennial underground runners give rise to coarse aerial stems. The conspicuous joints along each of these stems are an important identifying characteristic. The stems of the horsetail, *Equisetum arvense*, may grow to two feet and are branched at each joint in symmetrical whorls. The scouring rush, *Equisetum hyemale*, growing somewhat taller, maintains evergreen unbranched stems.

The horsetails experience a life cycle similar to the club mosses and ferns. Furthermore, the shoots, arising from the underground rhizome, may be either fertile (spore-bearing) or sterile. Both *E. arvense* and *E. hyemale* demonstrate this notable variation.

The scouring rush produces only one type of vegetative body. Pale green jointed stems are tipped with a spore-bearing spike. Spores are released throughout the summer months, but the evergreen stem lingers throughout the winter.

The common horsetail produces two types of vegetative stems. In early spring a rather feeble, flesh-colored stem supports a cone-shaped head of spores. Before the onset of the steady warmth of summer, these spores are freed and the fertile stems quickly die back. Shortly thereafter the sterile, pale green familiar horsetails replace the spent spore bearers.

The *Equisetum* species share another curious feature. The fibrous stems and branches are infused with minute silica crystals. The scouring rush, thus abundantly endowed, has enjoyed long service as a cleanser. Among both early people and modern recreational campers, many cooking vessels have been scrubbed clean with this natural scouring material.

Common Club Moss

Lycopodium clavatum

DESCRIPTION

Classification: Club moss
Size: Height to 10″
Characteristics: Green, ¼″ leaves densely packed around the stem; brown conelike spore cases held above the main stems

HABITAT

Moist soil in wetlands margins

RANGE

Maine to North Carolina

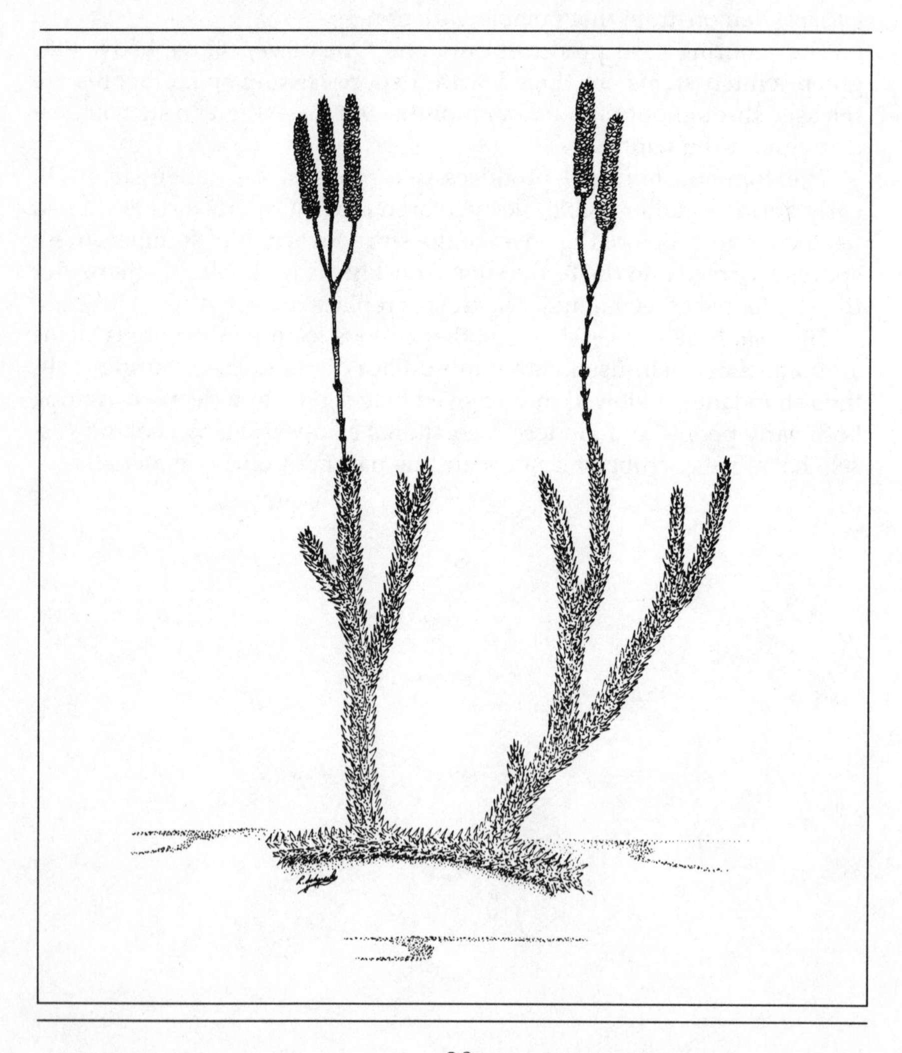

THIS humble little plant with origins in the ancient past enriches our lives in manners ranging from the exciting to the mundane. Its small size and bright evergreen leaves make it a popular Christmas decoration, and it is often overcollected, leaving an area with nothing but its memory.

The minute spores of the club moss, called *Lycopodium* powder, are used in the manufacture of pill coatings and bath powder and are a standard item in physics and chemistry classrooms. Very easily ignited, *Lycopodium* powder gained popularity as flash powder during the infancy of photography. The pyrotechnic industry is still somewhat dependent upon the powder in the manufacture of fireworks.

Associating the mysterious wetlands with the treachery of wolves, some envisioned the plant as a replica of a wolf's paw. Thus, the genus name *Lycopodium*, meaning "foot of the wolf," was adopted.

Club mosses form small colonies by spreading underground runners. Growing upward from the stem tips, thin leafless stalks fork and hold brown, conelike spore cases. These cases mature in late summer, and the spores are set free in great clouds.

Ground Pine

Lycopodium complanatum

DESCRIPTION

Classification: Club moss
Size: Height to 5″
Characteristics: Tiny, green, overlapping leaves held flat against the stem

HABITAT

Moist soil in wetlands margins

RANGE

Maine to Georgia

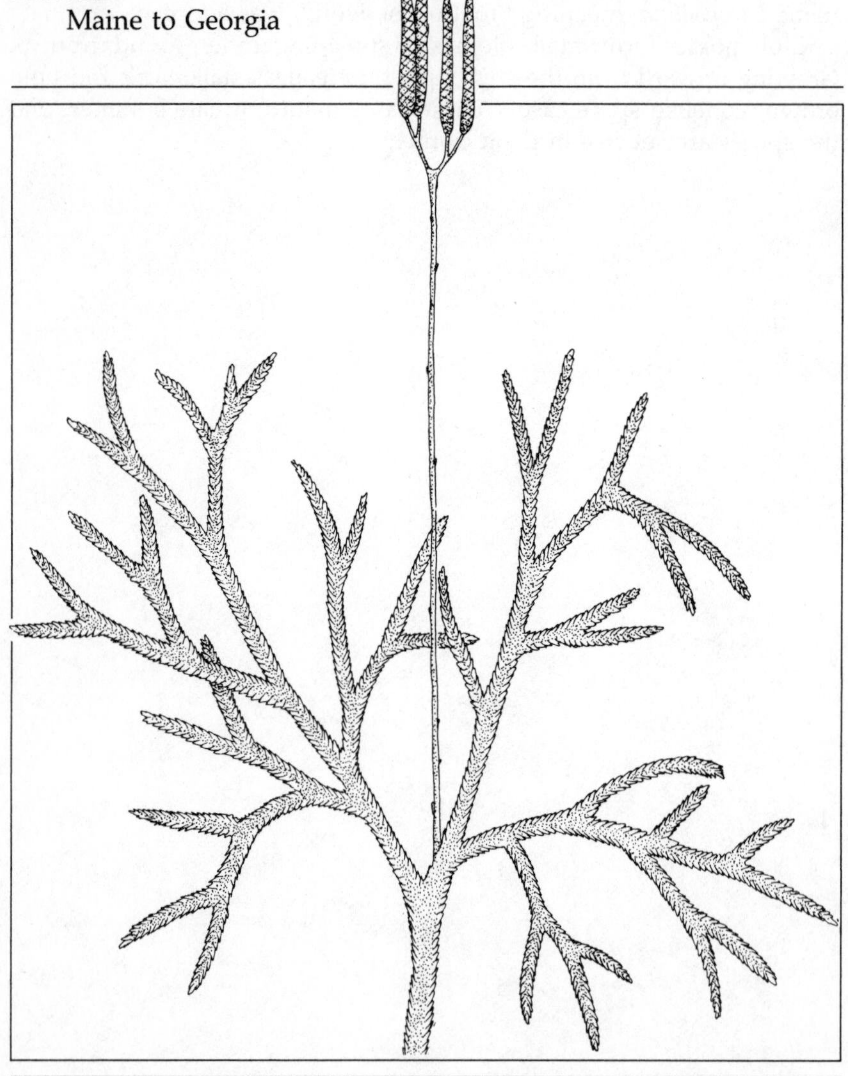

DURING the Paleozoic Era ancestral forms of the ground pine grew to heights of nearly sixty feet. These were the most successful plants of this time. Among the first to specialize in water-carrying tissue, the ground pines were the first land vegetation to develop true roots, leaves, and stems. As these trees died and were covered over with soil and new growth, the tremendous heat and compression brought about changes unique to the Carboniferous Period. Coupled with incomplete decomposition, the intense pressure upon these remains created the coal deposits that were eventually to power an industrial age. Even the compression of the spore accumulations formed a coal variety called cannel coal.

Today's ground pine, Lilliputian counterparts to the once massive forests, have been the subject of sacred Druid ceremonies and the idle curiosity of naturalists. Their small size and bright evergreen leaves have encouraged intense collection for holiday decorations. Appearing to be quite distinct and separate plants, many actually emanate from a single, underground runner. Therefore, an entire stand of these elfin pines may be unwittingly obliterated.

The life cycle of the ground pine, which undergoes the so-called alternation of generations, is similar to that of its moss and fern relatives. The spore-producing stage is much more prominent than the gamete-producing stage. The spores germinate in moist, warm soil to produce a tiny thallus, usually overlooked by the casual observer. Within the thallus, the sperm and egg, the germ cells of a new plant, develop. Thus, these unassuming plants persevere for yet another generation.

Reed Grass

Phragmites communis

DESCRIPTION

Classification: Grass
Size: Height to 12′
Characteristics: Long, narrow leaves alternating on tall stalk; plumelike flowers

HABITAT

Marshes

RANGE

Continental U.S.A.

A GENTLE breeze stirs a stand of reed grass into a restless flutter. The feathery plumes of flowers, held at the tip of the stiff yet yielding stalks, catch the slightest air currents and choreograph the wetlands' own "amber waves." Like the reeds themselves, the flowers are more likely to be observed en masse than singly. Wisps of silky hairs adjoin the tiny flowers into an elegant aigrette. Progressing through the season from tan, purple, and pink to fluffy white, the reed flowers are probably the most striking feature of the winter wetlands.

Phragmites has enjoyed a reputation as an agent of flood control and simultaneous notoriety as a pest. A prostrate stalk sends out runners to generate new plants. Stout rootstocks, often exceeding twenty feet in length, interlock to form a dense network. These sturdy soil anchors can withstand fires, mowing, and other forces that severely damage the stalks and leaves. Wildlife management often includes mechanically crushing the underground root system. This is done to prevent the natural tendency of the plant to seize control of a site and callously evict the native plants.

While the plant provides excellent shelter for wildlife, it is of little food value. Man has used the stalks and leaves extensively. In the Southwest, *Phragmites communis* is called "carrizo" and is used for lattices in the construction of adobe huts. Indians have used the plant for arrows, woven mats, screens, cordage, and roof thatching. Adorning the modern home, reed grass is often included in natural displays of dried flowers.

Wild Rice

Zizania aquatica

DESCRIPTION

Classification: Grass
Size: Height to 10′
Characteristics: Tall, emergent grass; light green, lance-shaped leaves; flower cluster with broomlike branches at the summit and dangling branches at the base; elongated ⅝″ fruit contained in papery husks; flowers June–August

HABITAT

Marshes

RANGE

Maine west to Wisconsin; south to Louisiana and Texas

SHIMMERING in the warm breeze, the tall, graceful, emergent wild rice annually charms the wetlands with its comeliness. Abundantly crowned with delicate flowers, this valuable food source intimately blends the aesthetic with the practical. Undulating to a soothing rhythm, it attracts an astonishing potpourri of game birds, songbirds, muskrat, deer, and moose.

Wild rice is a robust plant that often attains a height of ten feet. It has light green, backward-curving, lance-shaped leaves arising from a sheath near the base of the stem. Examination of the large, terminal panicle reveals that the uppermost areas contain erect female spikelets, while the lowermost areas contain the purplish, pendulous male spikelets. Upon pollination, papery, bristly tipped husks containing brownish, cylindrical fruits will form.

Wild rice has been cultivated by man to supplement the waterfowl food supply. It is usually planted by seeding in early spring or late fall. Soft, muddy areas that have gently circulating water make ideal sites. Ducks that have fed upon wild rice are reported to be superior food for man.

Wild rice is also used as food by man. It is grown and harvested commercially by American Indians in Minnesota. The ripening grains are collected just before they drop of their own accord. The grains are then dried thoroughly and rubbed gently to remove the husks. After being washed in cold water to subdue the smoky flavor, the grain is then prepared like brown rice or ground into flour. On rare occasions a highly poisonous pink fungus, ergot, *Claviceps* sp., can replace some of the seeds. When hardened, these fungi assume the seed's shape and size. If these fungi are present, the grain should be collected in another area.

Great Bulrush

Scirpus validus

DESCRIPTION

Classification: Sedge
Size: Height to 8′
Characteristics: Round green stem; single leaf at tip with loose cluster of green flowers

HABITAT

Marshes

RANGE

Continental U.S.A.

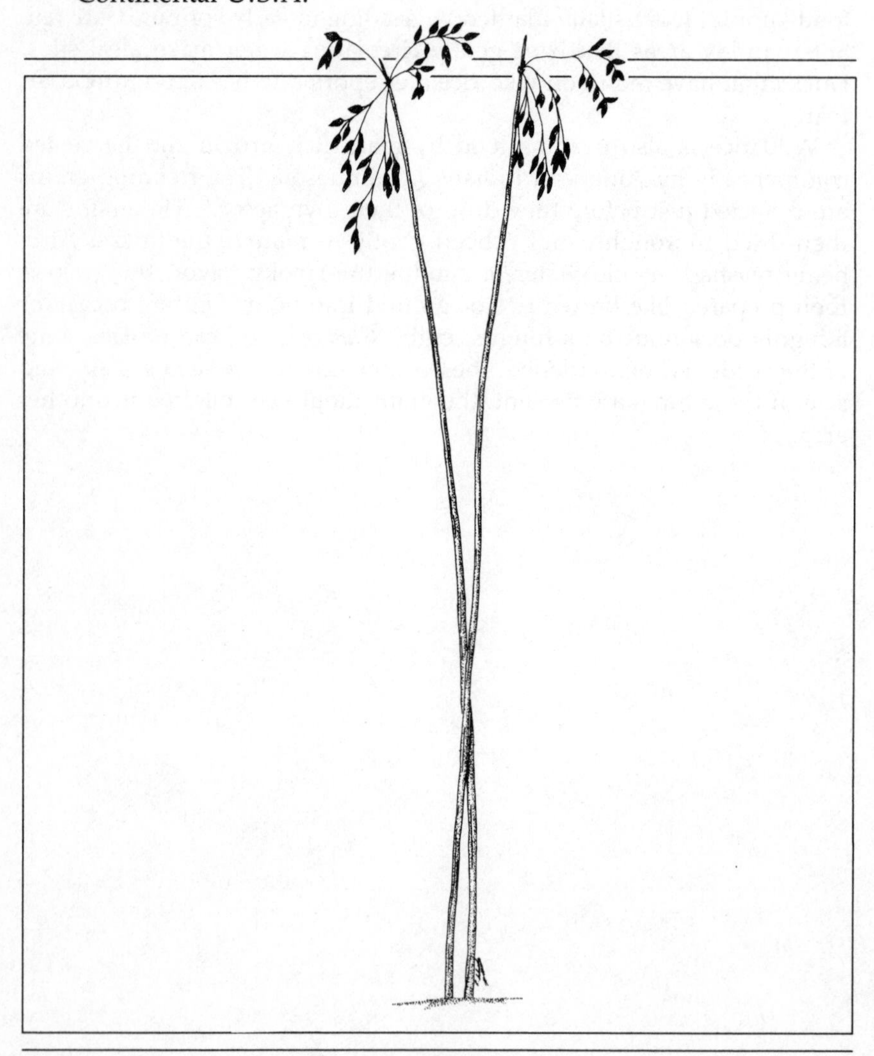

A WETLANDS naturalist who hesitates to attempt the deeper waters of a marsh may easily overlook the wildlife sanctuary provided by the great bulrush. A member of the sedge family, this simple perennial is an excellent indicator of the current character of a wetlands region.

Populations of most plant species vary considerably from year to year owing to rainfall. This single physical parameter most effectively determines the type and density of the dominant marsh vegetation and the corresponding animals occupying that ecological niche.

The great bulrush, an emergent reed plant, generally flourishes in several feet of water. Thick stands, intermixed with cattails, offer excellent shelter, food, and nesting sites for mammals and waterfowl that prefer aquatic lifestyles. The timid least bittern builds her nest over water, often on reeds of great bulrush. Here she can raise her brood, providing them with a variety of fish and frogs, while studiously avoiding dry land.

Muskrats also profit from the great bulrush. Like many species of ducks, these aquatic rodents enjoy the nutritious tuberous rootstocks year round. The long, round reeds are often interwoven with other vegetation as muskrats engineer their elaborate lodges. Stripped and shredded, the reeds are also an important bedding material for their nests.

The very tip of the bulrush reed gives rise to two structures, the leaf and the flower cluster. The single, thin, bladelike leaf arches gently downward, demurely deferring to the dainty flowers. Each miniature blossom crowns a long, erect petiole, forming a loose cluster.

Neither the flowers nor the nutlets that they issue are the principal means of reproduction. Instead, fed by the photosynthetic activity within the stem and leaf, the tuberous rootstocks grow and spread, generating more and more shoots each spring. This interlocking mass of tubers, densely matted in the saturated earth, is an excellent soil anchor. Thus bulrushes are well regarded as land builders.

Man has also enjoyed the gourmet qualities of the great bulrush. Cooked and dried, the roots can be ground into a sweet flour. The young, pale shoots have been lauded as a tasty treat either raw or cooked.

Historically the great bulrush has been an important natural resource. Mats, beehives, baskets, and chair seats are among the more popular items woven from the stems of this plant.

Several other species share the common name bulrush. Particularly, certain grasses of the genus *Juncus* are often called bulrushes. The biblical bulrushes used to cradle the infant Moses were actually the waterside sedge, papyrus. Such misnomers, however, cannot diminish the countless contributions of the great bulrush, *Scirpus validus*, to the wetlands community.

Spike Rush

Eleocharis sp.

DESCRIPTION

Classification: Sedge
Size: Height to 5′
Characteristics: Round, leafless stalks tipped with a flower cluster

HABITAT

Marshes

RANGE

Continental U.S.A.

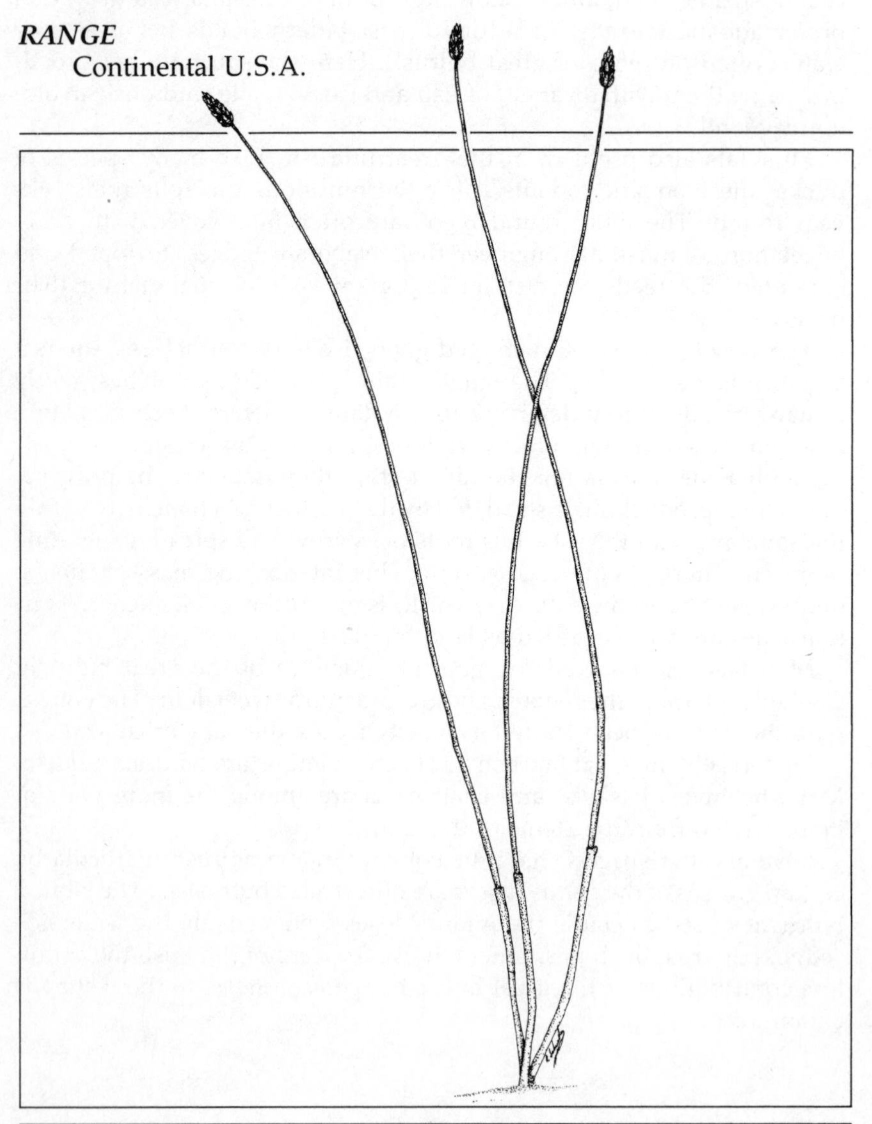

THICK bouquets of emerald canes rise conspicuously above the water's surface. The leafless stalks of a spike rush cluster stretch skyward, each holding a conical mass of flowers at its tip.

Although they are sedges and not true rushes, the many species of *Eleocharis* form a group known collectively and individually as spike rushes. These plants are most comfortable when rooted in well-saturated soil. Varying in size from less than a foot to nearly five feet, most species are commonly encountered along wetlands margins or even partially submerged within the main body of a marsh. The genus name is a derivative of *elos*, the Greek word for marsh.

Since the genus is characterized by round, leafless stems, species identification usually relies upon the spikelet, or flower cluster. Each stalk is tipped with miniature flowers arranged in a spiral. The tiny fruit structure, forming after pollination, varies considerably from species to species.

Spike rushes grow from horizontal underground stems that form a matted rootstock. Although not easily separated, the rootstock is usually subjected to artificial divisions to propagate the plant where populations of *Eleocharis* are desirable.

Stands of spike rushes make excellent cover for a variety of wetlands species. The nutlets and stems are popular food items for waterfowl, marsh birds, and muskrats. The tasty water chestnuts in Chinese foods are cut from the tubers of the Chinese spike rush, *E. tuberosa*.

Tussock Sedge

Carex stricta

DESCRIPTION

Classification: Sedge
Size: Height to 3′
Characteristics: Narrow, sharp-edged leaves forming dense clumps

HABITAT

Swamps, marshes

RANGE

Northern U.S.A.

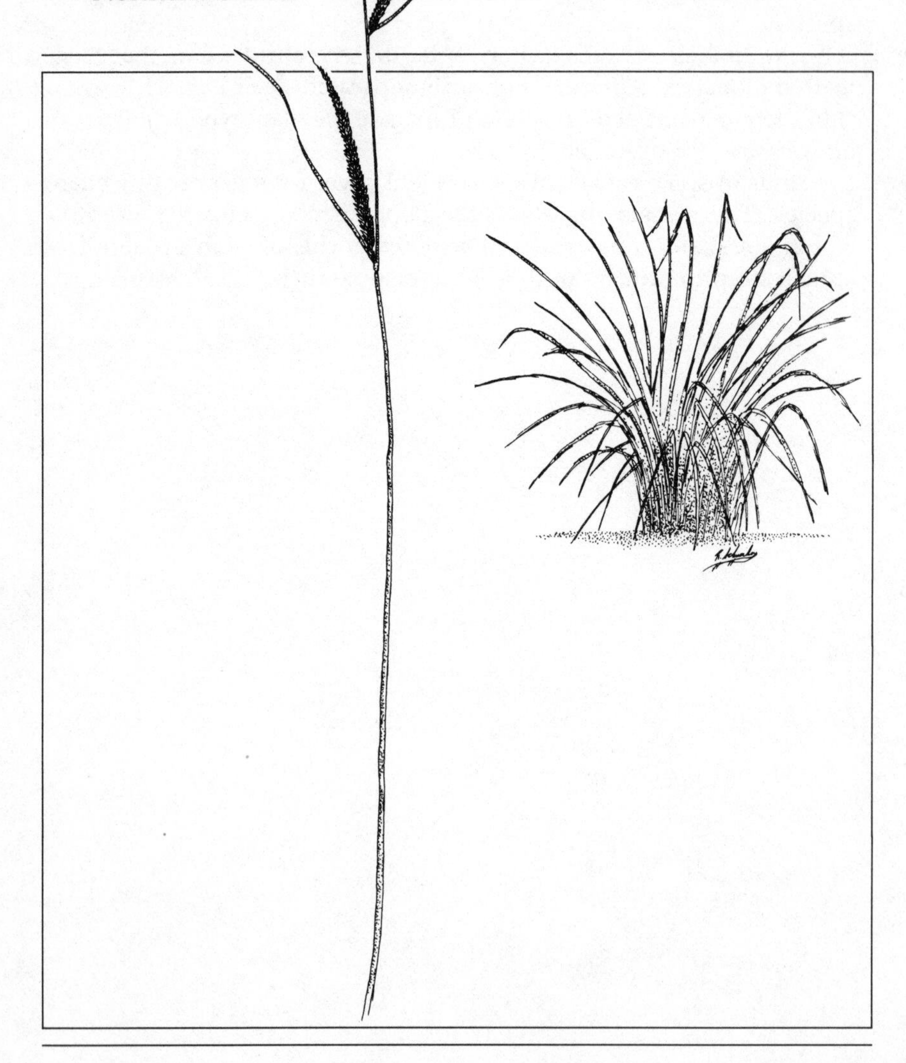

CLUSTERS of *Carex* sp. are prominent in wetlands across the United States. Dotting the busy landscape, each tussock emerges from the water as a mini-island. Indeed, the unprepared naturalist may attempt to negotiate a marsh by leaping from cluster to cluster, only to slip from the dew-glazed clump and meet the water firsthand.

The various species of sedges usually grow in pure stands. The soft, flat leaves spring from triangular stems and arch gracefully. The leaves on a tussock's perimeter often curve backward to meet the water's surface. Spikes bearing flowers are numerous in a single plant cluster. Pollination is for the most part effected by wind, although insect activity contributes to a small degree. The fruit, called a nutlet, is surrounded by a thin, brittle sac.

Sedges perform several vital functions in a wetlands habitat. A service of immediate importance to man is soil building. The well-developed root system anchors the soil in an environment plagued with fluctuations in the water level. Thus, erosion is slowed as the effects of spring flooding, summer droughts, and autumn rains chafe the soil line.

Wetlands wildlife is also well served by sedges. They provide an excellent cover for small animals hiding within the safety of the dense foliage; they provide a welcome perch for frogs and turtles sunning their sluggish forms; they provide diverse forage for beaver, muskrat, deer, waterfowl, and a variety of birds. These animals may feed on the leaves, roots, young shoots, or nutlets.

A tussock sedge cluster, representing an area of drier soil above the water's surface, may also befriend an errant seed or spore ripe for germination. Thus, a small patch of a single species, by soil building and sharing its habitat, may soon nurture an exotic bouquet of vegetation and its attendant animal life.

Arrow Arum

Peltandra virginica

DESCRIPTION

Classification: Wildflower
Size: Height to 18″
Characteristics: Large, arrow-shaped leaves; slender spadix almost completely concealed by an erect spathe; flowers May–July

HABITAT

Swamps, bogs

RANGE

Maine to Michigan; south to Florida, Louisiana, Missouri

THE brilliant green arrow arum is common in low, flooded areas. It unfurls splendidly in the early spring into a heavily veined blade resembling a rather stocky spearhead. As the plant matures, it forms a white fingerlike spadix that is almost completely concealed by a four- to eight-inch erect, pointed spathe. Bearing male flowers at the top and female flowers at the base, the spadix is protected by the enveloping spathe. When the plant is fertilized, green berries with one to three seeds each are formed within the spathe. A related species, *Calla palustris*, has a broader white spathe and produces crimson berries.

Wood ducks are probably the only large animal to enjoy the fruit of the arrow arum. The flowers, however, attract a wide range of insects, making the plant a favored fast-food emporium of many amphibians that delight in a diversified diet.

Throughout our history, a variety of recipes have offered the arrow arum integrity as a food item. Processing may require more energy, however, than the nutritional value can justify. Containing caustic calcium oxalate crystals, the rootstock must be carefully prepared to avoid severe internal distress. Boiling alone does not remove this property. The arrow arum must be roasted for several hours and then dried for six months before being ground into flour.

Broad-Leaved Arrowhead

Sagittaria latifolia

DESCRIPTION

Classification: Wildflower

Size: Height to 3′

Characteristics: Arrowhead-shaped leaves; white flowers in whorls of three on erect stems; flowers July–September

HABITAT

Swamps, marshes

RANGE

Continental U.S.A.

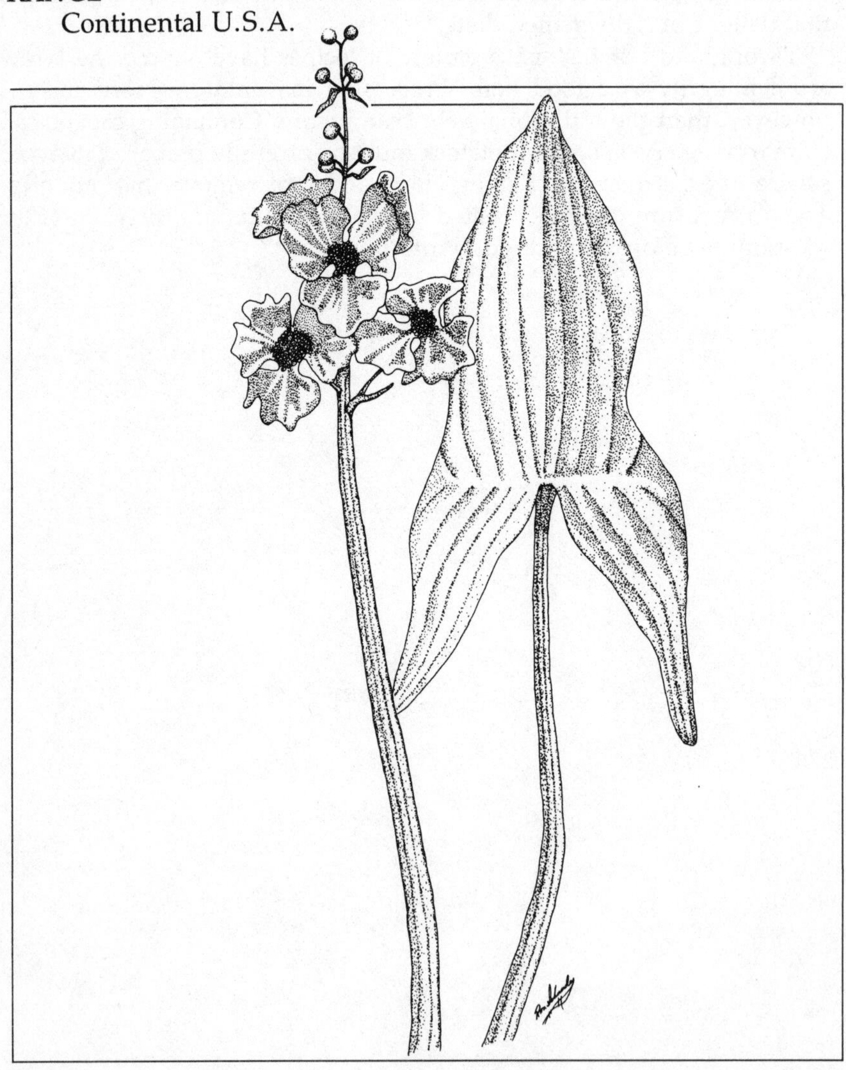

A MARSH whose edges are bordered with the delicate arrowhead offers a charming diversion to the weary naturalist. Colorful wood ducks and canvasbacks may be seen adeptly navigating through the blossoms seeking the delicacies rooted below.

This wildflower is very abundant and well deserving of its name. The leaves held above the water are shaped decidedly like an exaggerated arrowhead. This same plant, however, has long, narrow, underwater leaves, similar to grass blades, which can withstand the water currents. The snowy white petals of the male flowers grow in whorls of three on the upper portion of the stem, while the inconspicuous, dull female flowers spiral on the lower stem.

Growing to a height of three feet, the arrowhead is a striking, succulent perennial. Since transpiration through the leaves is extensive, these plants are often eliminated in reservoirs to minimize water loss. Thus, once again, man sacrifices the aesthetic for the practical.

The broad-leaved arrowhead is often called the swamp or duck potato because of the potato-like starchy tubers along the root system. Ducks, geese, swans, and muskrat feed on these, although it takes much strength to dig them up.

Wading American Indian squaws located the duck potatoes with their toes. When loosened with long sticks, the potatoes floated to the water's surface. Although slightly unpleasant raw, they are flavorful when cooked prepared as one would potatoes.

Blue Flag
Iris versicolor

DESCRIPTION

Classification: Wildflower
Size: Height to 3′
Characteristics: Swordlike leaves arising from a basal cluster; showy, violet-blue flowers; stiff, rodlike stem; flowers May–August

HABITAT

Swamps, marshes

RANGE

Maine to Virginia; west to Ohio, Michigan, Wisconsin and Minnesota

THE iris has been prominent throughout history. Pliny writes of instructions for the ceremonial gathering of iris roots and of the medicinal properties of the plant. To the ancient Egyptians, the iris was a symbol of eloquence and was used to adorn the brows of the sphinx statues. In France, chosen because of its exquisite yet statuesque beauty, the iris became the emblem of the house of Louis VII. It was given the common name of fleur-de-lis, meaning the flower of Louis, *lis* being a corruption of the name Louis.

Today, this colorful plant is abundant, both as a cultivated garden favorite and as a captivating wildflower. Displaying many varied hues, the flower was named in honor of Iris, the goddess of rainbows.

The blue flag, *Iris versicolor*, one member of the extensive iris family, is particularly fond of wet areas. A similar species, *Iris virginica*, is slightly smaller in size and may be found in southern wetlands. Forming a thick, purple parcel, a profusion of these lofty flowers provides an exciting splash of royal color to a marsh.

This herbaceous perennial arises from a thick, fleshy, horizontal rhizome that is fibrous and light-colored. The leaves, emanating from a basal cluster, are swordlike, shorter than the stem, and erect. The two- to three-inch flowers of the blue flag are violet-blue and are tinged with yellow or white. Each consists of three large violet outer sepals, all prominently veined and downward curving, with one being bearded. The three smaller erect inner parts are the true petals, which unite to form a short tube at the base.

The unique shape of the iris flower represents a remarkable adaptation of the plant to ensure cross-pollination. Since the position of the stamens is such that their pollen could not easily reach the stigmas of the same flower, self-fertilization is not easily achieved. On the other hand, cross-pollination is encouraged by several factors. Bees are strongly attracted to blue or violet flowers, especially those that are large and showy, and the iris fills this requirement well. The structure of the iris flower tends to direct the bee downward toward the nectar. In pursuit of this tasty treat, the bee is dusted with pollen, which is then carried to another flower.

The rootstock of the iris contains irisin, which can cause indigestion in humans and animals. Since the roots have a very offensive taste, however, ingestion is rare, although the dried roots have been used as a cathartic agent. Severe irritation of the skin may also be caused by handling iris roots. Orris root, which comes from the *Iris florentina* and smells like violets, is used as a perfume and a flavoring.

The leaves of the blue flag provide some shelter for aquatic game. Overall, however, the plant is of little food value to most wildlife.

Marsh Buttercup
Ranunculus septentrionalis

DESCRIPTION

Classification: Wildflower
Size: Height to 3′
Characteristics: Glossy yellow, five-petaled flowers, 1″ in diameter; many-branched stem; three-parted leaves; flowers April–July

HABITAT

Swamps, marshes

RANGE

Maine to Georgia; west to Kentucky and Missouri

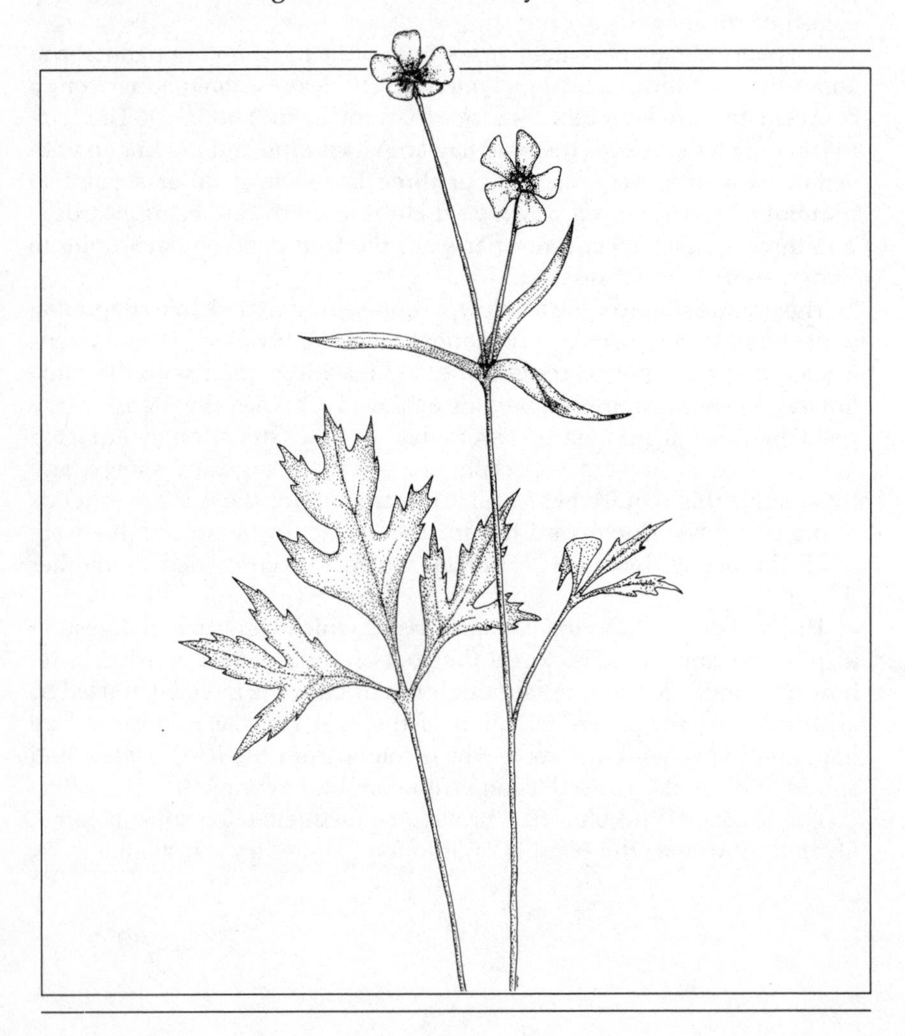

AS the insulating snow blanket of the swamp begins to vanish, the underlying earth is once again exposed to the warming influence of the sun. Releasing a rich, damp earthy odor, the thawing soil becomes dotted with the glossy golden chalices of the marsh buttercup. Shining brightly in the early spring sunlight, these delicate flowers stir gently in the balmy breeze.

A favorite of children, the marsh buttercup is happily plucked and held under the chin of a companion. If a yellow reflection appears on the child's face, it is naturally assumed that the playmate is fond of butter.

The buttercup, which tends to grow in wet places, has the genus name of *Ranunculus*, which is derived from the Latin word meaning "little frog." However, of the thirty species known, there are some that prefer fields and woods.

The many-branched, fuzzy stem of the marsh buttercup is both upright and trailing. The compound leaves are unequally three parted, with the divisions being stalked. The yellow petals form a cuplike flower that is designed to prevent self-pollination. Since the anthers open away from the pistils and shed most of their pollen before the stigmas are ready to receive the grains, self-fertilization is limited. Cross-pollination is accomplished largely by beelike flies of the family Bombylidae and by the small bees of the family Andrenidae. After pollination the flowers produce a tiny cluster of dry fruits called achenes. Each achene has a winged margin and a birdlike beak.

The juice of both the stem and leaves of the marsh buttercup may cause severe dermatitis. If eaten, the plant may also cause intense gastric disturbances, and for this reason the marsh buttercup is usually avoided by foraging animals.

Cardinal Flower

Lobelia cardinalis

DESCRIPTION

Classification: Wildflower
Size: Height to 4½′
Characteristics: Brilliant scarlet flowers; stiff, unbranched stems with alternating leaves; flowers July–September

HABITAT

Marshes

RANGE

Maine to Florida; west to Texas and Colorado

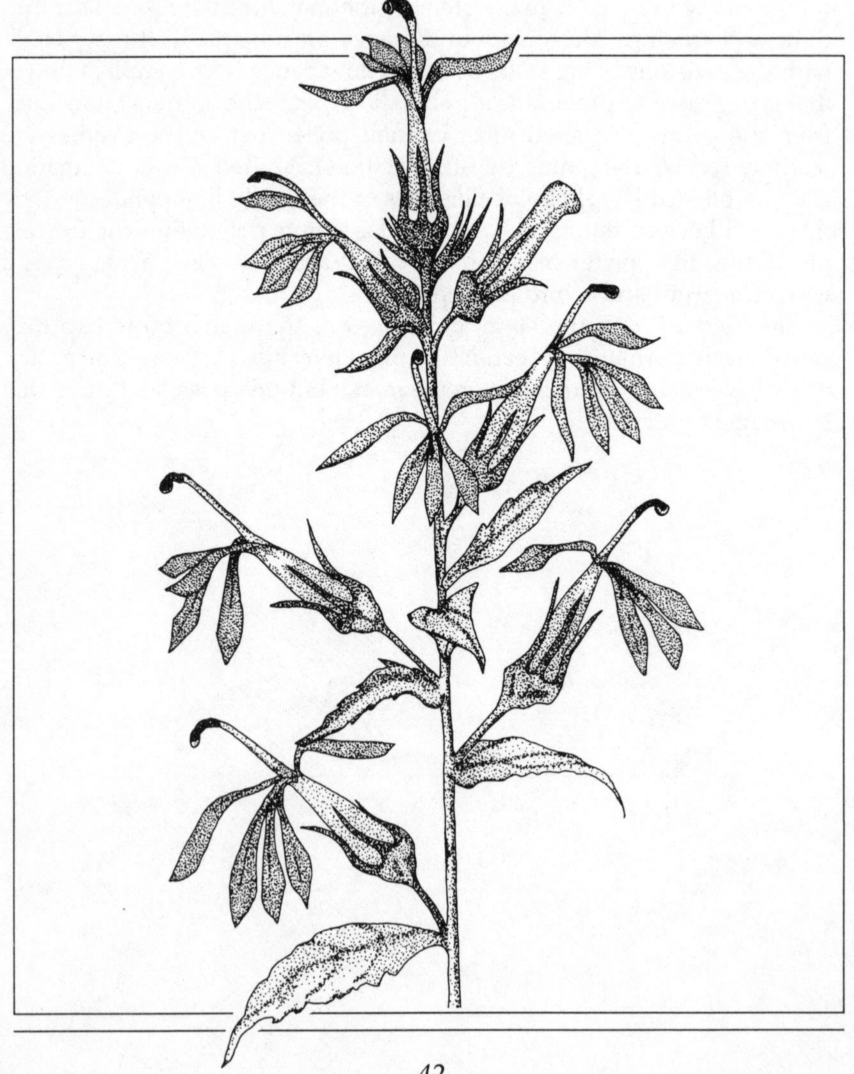

ON a sultry day in late summer when the sound of the cicada hangs thickly in the air, a fortunate few may chance upon a cardinal flower. Rising among the tangled green hues of the marsh, the vivid blood red flower delights both the eye and the spirit.

The cardinal flower is indigenous to America, and about thirty different species grow here. Early French explorers found the plant so captivating that they sent it back to France as an elegant sample of the New World. With blossoms suggesting the rich, red robes of the Catholic cardinals, perhaps it was then that the plant received its name.

Standing stiffly erect and unbranched, the stem is well supplied with alternating leaves. The blossoms are borne in clusters at the stem tips. The tube-shaped flowers have two lips, the upper having two lobes and the lower having three lobes. Since the flower tube is fairly long and the lower lip so fragile, many insects cannot land on the plant to reach the nectar. Thus, pollination is accomplished chiefly by hummingbirds. The hummingbirds do not need to alight, and their long tongues are adapted for easily reaching the nectar deep within the flower. After pollination a pod with many seeds is formed.

Containing powerful alkaloids, the entire plant was valued by the Indians for medicinal purposes. However, when used in excess the curative powers of the cardinal flower often proved deadly.

The cardinal flower is not hardy enough to withstand man's onslaught. It has suffered in recent years from zealous overcollection and habitat destruction. While many marsh species have recovered from such trauma, the cardinal flower may become but a faded memory of the wetlands naturalist.

Broad-Leaved Cattail

Typha latifolia

DESCRIPTION

Classification: Wildflower
Size: Height to 8′
Characteristics: Stiff stems with brown, cigar-shaped "cats' tails" at the tips; alternate, pale green, ribbon-shaped leaves; flowers May–July

HABITAT

Swamps, marshes

RANGE

Continental U.S.A.

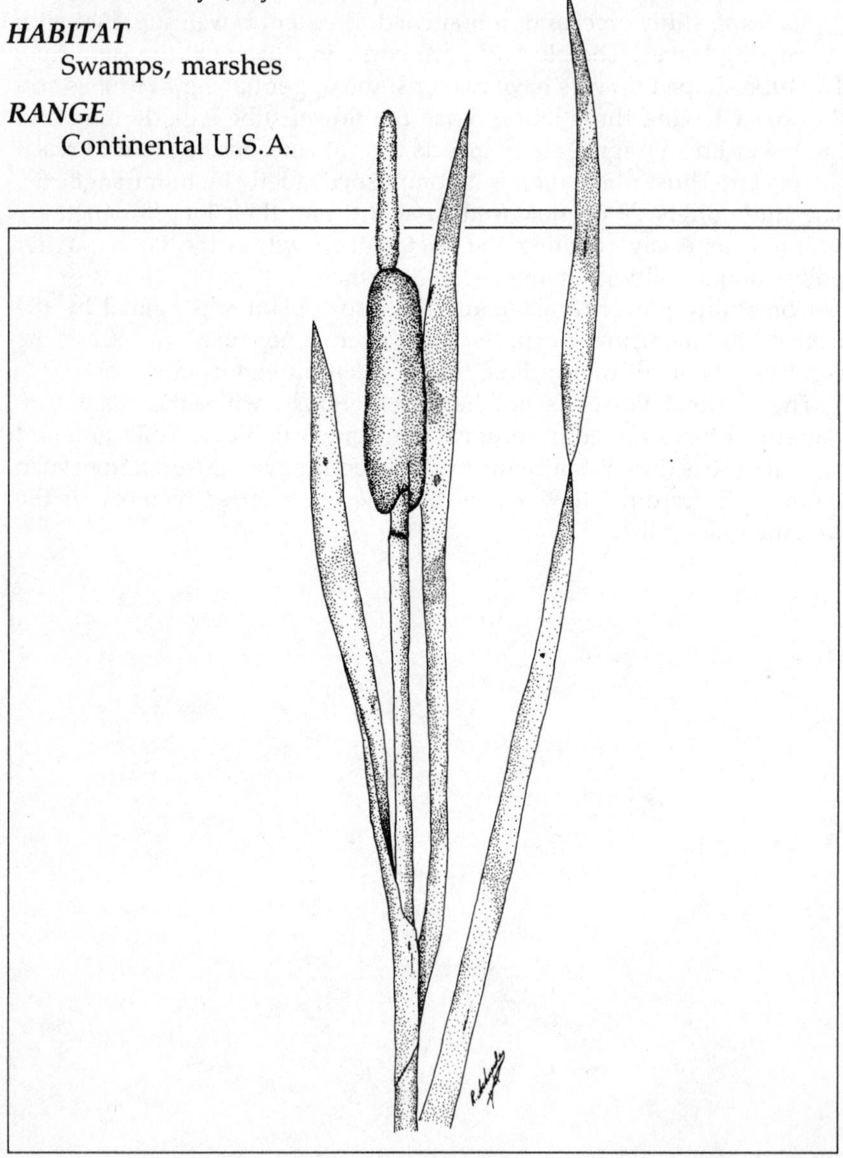

THE misty mornings of late autumn soften the austere landscape of the muted marsh. The first red rays of the sun peeking over the frosty horizon set the crystalline cattails ablaze. These lonely sentinels keep a vigil over the slumbering denizens of their domain.

The familiar cattail, often forming extensive stands in marshy areas, is a key plant in changing wet soil into dry land. The thick mat of stocky, horizontally branching rootstocks traps decaying debris. As this debris is broken down to form soil, the cattails slowly succumb to succession.

During the summer, the cylindrical cattail is composed of thousands of minute flowers. The upper, light-colored flowers are staminate, while the lower, darker flowers are pistillate. A short gap of bare stem between the male and female flowers is evident in the narrow-leaved cattail, *T. angustifolia,* but not in the broad-leaved cattail.

After the dispersal of the pollen, a withered upper stalk remains. The fertilized female flowers bear small, seedlike fruits with long hairs growing from their stems. All through the winter, the cattail bursts into a fluffy mass releasing the fruits to the winds.

Although providing excellent shelter for waterfowl and a number of songbirds, cattails offer little food value to these creatures. On the other hand, the muskrat enjoys feeding on the cattail shoots in the spring, the leaves and stems in the summer, and the roots throughout the fall and winter. The muskrat also finds safety in the dense growth and uses the stalks and leaves to build its home.

The cattail moth lives primarily in the cattail marshes, hibernating as larvae in the seed head of the cattail. Many of the fluffy seed heads seen throughout the winter are inhabited by the moth larvae which use their silk to bind the seeds to the stalk.

The American Indians, who had high regard for the cattails, ate the roots of the young plants, whether cooked or raw, and made porridge of the seeds. The silky seed hairs were used as an absorbent material in the Indian's papoose carriers.

The cattail was once used to produce paper. Today new technology is attempting to refine this paper production process for modern use. The leaves have long enjoyed prominence in the manufacture of rush-bottomed furniture. While the plant had been used for caulking barrels, this practice is now obsolete.

Cattails are a versatile food plant. The edible fruits forming at the tip of the cattail are called Cossack asparagus. The young shoots and stems may be eaten cooked or raw. Like most vegetables smothered in butter, the immature flower spikes can be quite tasty. Pollen mixed with the ground rootstocks produces a protein-rich flour. Freshly baked sweet rolls served warm with wild strawberry jam are a tasty holiday treat.

Marsh Cinquefoil

Potentilla palustris

DESCRIPTION

Classification: Wildflower
Size: Height to 2′
Characteristics: Deep purple blossoms with short, narrow petals and wide purple sepals; sprawling stem; pinnately compound leaves; flowers June–August

HABITAT

Swamps, marshes, bogs

RANGE

New Jersey west to Pennsylvania; Ohio; Indiana; Illinois; Minnesota

FOUND throughout wet marshy areas and bogs, the tangled, sprawling stems of the marsh cinquefoil abound with tiny amethystine blossoms. Often overlooked because of their small size, the flowers first appear in early summer and bear a distinct resemblance to the wild rose.

The name cinquefoil is derived from two words: *cinque*, meaning five, and *feuilles*, meaning leaves. Of the three hundred species of cinquefoils known, the most common is the five-fingered cinquefoil, referring to its five leaflets that resemble the spreading fingers of a human hand.

The marsh cinquefoil can attain a height of two feet and has reddish colored, trailing stems that are usually rooted in water or mud. The deep green, oblong leaflets are sharply toothed and are found alternating along the downy stems.

The flowers of the marsh cinquefoil stand erect and have five small, pointed petals that are narrower than the five broad, purple sepals. After pollination by minute flies, small dry fruits are produced.

Pliny wrote that cinquefoil mixed with honey and grease formed a salve that was used to treat a cutaneous form of tuberculosis. During the Middle Ages cinquefoils were regarded as having healing properties for almost any ailment and were widely collected. Even Linnaeus named the genus of the cinquefoils *Potentilla* from the Latin *potens* or "powerful," after their reputation as powerful cure-alls. Thus we are reminded today of the plant's therapeutic role in the history of man.

Cranberry

Vaccinium sp.

DESCRIPTION

Classification: Wildflower
Size: Creeping vine with branches 6–8″ high
Characteristics: Smooth, alternate, dark green, glossy leaves; nodding clusters of pink flowers; bright red berries; flowers June–August

HABITAT

Bogs

RANGE

Maine to North Carolina; west to Arkansas

A CRANBERRY bog, dressed in festive green and red, kindles the warm feelings of the Christmas holidays. The crimson berries were long used for food by early man and still provide us with such favorites as cranberry sauce, pie, juice, jellies, and jams. The tartness of the cranberry has given this plant the alternate name of sourberry. Therefore, it is customary to add a sweetener to make the fruit palatable.

The cranberry is a slender, creeping vinelike plant with tiny, leathery evergreen leaves. There are two species of cranberry, and they both flower in early summer. The wild species, *V. oxycoccus*, holds its pink, bell-like flowers at the tips of its stems, while the flowers of *V. macrocarpon* are scattered along the stem. The pollen, far too heavy to depend upon airborne pollination, is carried by bumblebees and honeybees. The fruit is set and ripens by late autumn. The larger berry of the *V. macrocarpon* is commercially marketed.

By Thanksgiving the cranberry industry will provide 20,000 tons of fruit for the holiday market. The cultivation of cranberries involves close regulation of the bog habitat. Certain native plants, such as the bog aster, tend to crowd out the more timid cranberry and must be culled in the name of commercial interest.

The water level of the bog is artificially controlled to enhance pollination, harvesting, and winter protection. After the spring thaw the water level is lowered to permit flowering and pollination. Once the berries have ripened, the bog is again flooded for the harvest. Specialized machinery shakes loose the berries and gathers them as they float on the water's surface. After the full crop has been collected, the water level is again adjusted. Securely encased within the ice, the vines are sheltered from the sharp winter winds and frigid air temperatures as they await the coming spring.

The need for high-energy, low-weight travel food is as old as man himself. The Indians processed dried strips of venison with ground cranberries to make a snack called pemmican. This was carried on long journeys.

Since cranberries have a slow spoilage factor and a high vitamin C content, they became standard fare on long sea journeys, where they were eaten to prevent scurvy. Our American heritage has been enriched by this versatile shining jewel of the bog.

Duckweed

Lemna sp.

DESCRIPTION

Classification: Wildflower

Size: ⅕″ diameter

Characteristics: Green, flat, oval floating plants

HABITAT

Swamps, marshes

RANGE

Continental U.S.A.

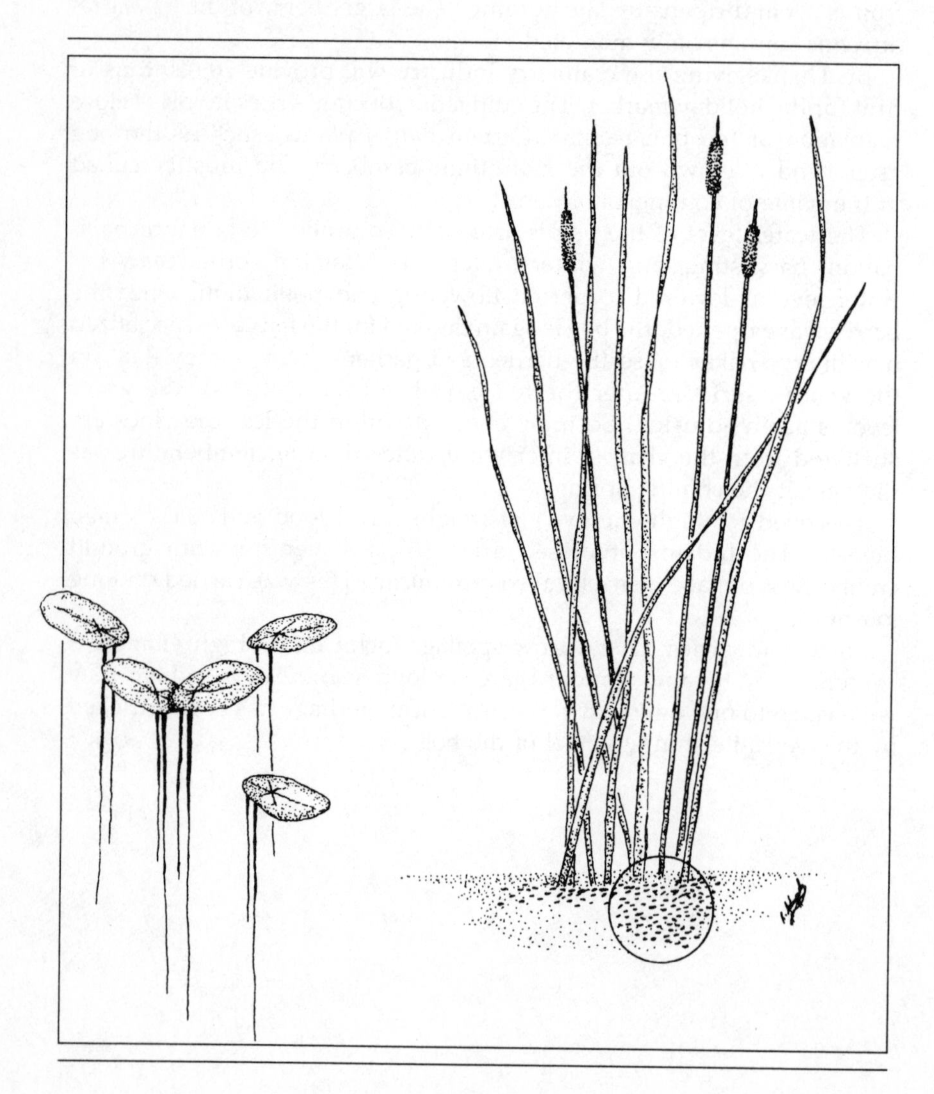

THE quiet waters of the swamps are covered with a dense, pale green carpet. As an unexpected form rises from the depths, the once smooth surface is broken into patches of a living mosaic. A head appears, a snake or perhaps a turtle, tiny plants clinging to its face. The seemingly uniform mat is actually hundreds of tiny, floating plants. Each is a separate organism, no larger than one-fifth inch in diameter, with a single dangling root.

Duckweed's ability to multiply rapidly is quite remarkable. Although these are among the smallest flowering plants, vegetative division is the common method of reproduction. Minute, inconspicuous flowers infrequently blossom and form tiny, thin-walled fruits.

Growing in association with the duckweed is watermeal, *Wolffia* sp. More granular in appearance than duckweed, watermeal is similar in structure but is a fraction of its size. A glimpse of the microscopic flowers is quite extraordinary.

Duckweed has a high nutritive value which exceeds that of most agricultural plants. Grazing eagerly on fields of duckweed, geese, ducks, and muskrats may obtain up to 90 percent of their dietary needs. Its exceptional food value, its camouflage factor for small prey, and its filtering effects on sunlight make duckweed a significant component of these wetlands habitats.

Golden Club
Orontium aquaticum

DESCRIPTION

Classification: Wildflower
Size: Height to 2′
Characteristics: Oblong leaves; minute yellow flowers borne on a single erect club; flowers April–June

HABITAT

Swamps, marshes, bogs

RANGE

Massachusetts to Florida

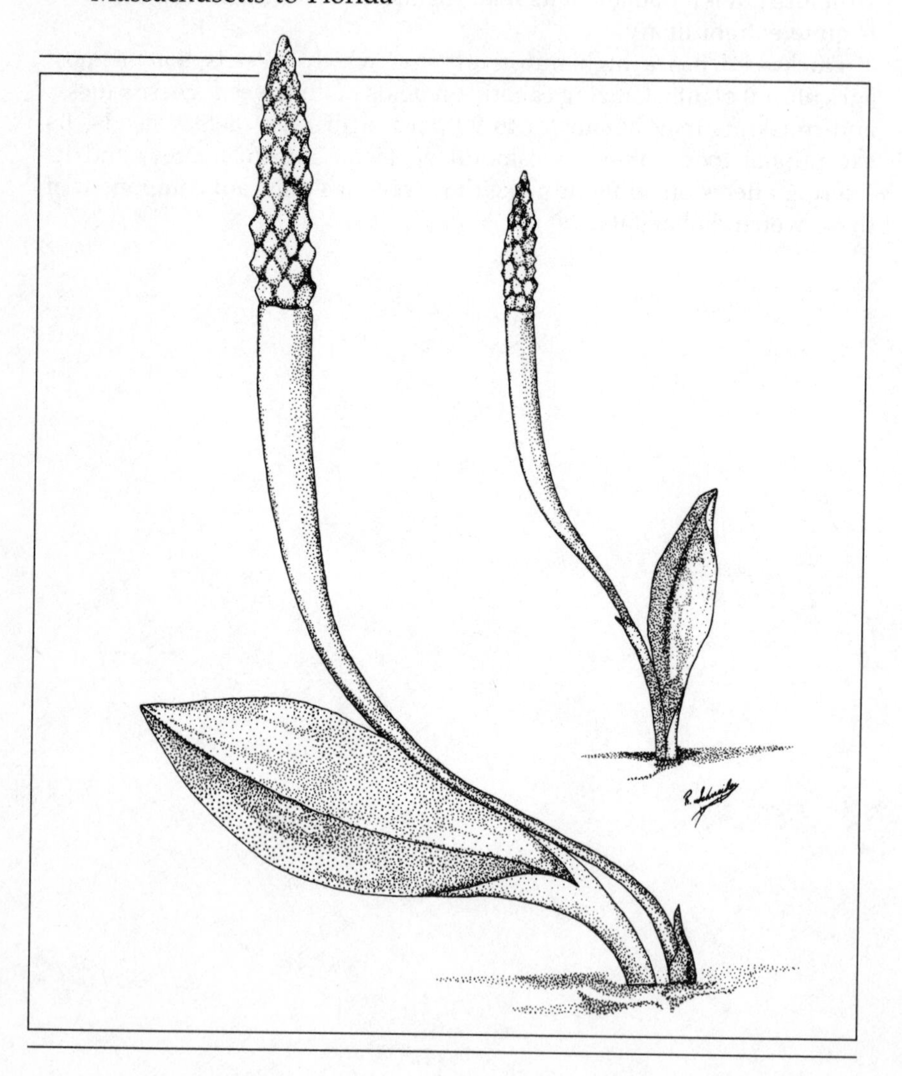

AN expedition into a secluded swamp on an early spring day, as the sunlight begins to strengthen and thaw the frozen earth, may reveal an unexpected surprise. Amid the glistening rivulets of water coursing merrily over, under, and around chunks of ice in the swamp, the thickened stems of the golden club stand firmly. Unfurling slowly, the dull green blades seem to challenge the chilling winds that gust around them, as they rise from the rootstock buried deep within the rich mud below. Dotted with hundreds of minute yellow flowers, the pale greenish white stem stands proudly, proclaiming, in its glory of color, the promise of the summer warmth yet to come.

A perennial aquatic herb, the golden club has long, stalked leaves that often float on the water. A thick, fingerlike golden spadix, located at the end of a fleshy, pale white stalk, is covered with tiny yellow flowers containing both stamens and pistils. The fruit produced by the plant is a blue-green bladderlike structure similar to green peas.

The golden club derives its scientific name from the Orontes, a river in Syria, where it was first discovered growing in rich profusion along the marshy edges. Its common name is a product of the crowded golden flowers that give a clublike appearance to the stem.

Being quite showy, the golden club is often grown as an ornamental in landscaped pools where it is started by cuttings. In the wild it can form a solid mass that extends for acres and may become a serious hazard to boat navigation.

The bulbous rootstock and blue-green seeds of the golden club, if thoroughly dried, can be ground into flour and used in baking. The seeds can also be eaten as a cooked vegetable if they are boiled for one hour in several changes of water.

Swamp Hellebore

Veratrum viride

DESCRIPTION

Classification: Wildflower
Size: Height to 6′
Characteristics: Stout, unbranched stem; large, heavily ribbed, clasping leaves; clusters of green, star-shaped flowers; flowers May–July

HABITAT

Swamps, marshes

RANGE

Maine to Georgia; west to Minnesota

IN early spring the somber gray-brown hue of the marsh is dotted with an array of tiny, bright green spears. Within a few weeks as the weather warms, these rigid miniature totems unfurl to form conspicuous masses of vivid green foliage. Often attaining a height of six feet, the swamp hellebore displays a coarse beauty. The mature plant flaunts a majestic presence in an otherwise bleak landscape.

The swamp hellebore is also known as Indian poke or false hellebore. It is a perennial and has a short rootstock with many thick, fibrous roots. It quickly succumbs to man's agricultural practices, since it cannot survive the drainage of its habitat.

The flowers of the swamp hellebore are displayed in clusters on a panicle located at the top of the plant stem. Yellow-green and star-shaped, each flower will produce a thick, smooth, one-inch-long fruit containing many seeds.

The swamp hellebore does not provide food for animals since all parts of it are poisonous. Containing a mixture of alkaloids such as veratrine, jervine, veratridine, and cevadine, it has a foul taste and is not normally enticing to wildlife. Accidental ingestion, however, is likely to result in vomiting, abdominal pain, tremors, and spasms. If the condition is left untreated, death may result.

Water Hemlock

Cicuta maculata

DESCRIPTION

Classification: Wildflower
Size: Height to 6′
Characteristics: Smooth, jointed stems streaked with purple; alternate, finely cut, compound leaves; umbrellalike clusters of white blossoms; flowers June–September

HABITAT

Swamps, marshes

RANGE

Maine to Florida; west to Texas; north to Minnesota

THE dainty, white flowers of the water hemlock are quite attractive and add a touch of finery to the countryside. Blossoming in midsummer, the water hemlock remains alluring until the nipping frosts of September come to claim it. Despite its seductive beauty, the water hemlock is a plant to be avoided. Containing perhaps one of the most dangerous vegetable poisons native to our country, the water hemlock may prove fatal if ingested.

A perennial herb with smooth, hollow, jointed, magenta stems, the water hemlock can attain a height of six feet. The reddish-tinged compound leaves are arranged in an alternate pattern on the many-branched stems, with the upper leaves being smaller than the lower ones. Containing many tiny, five-petaled blossoms, the umbel, an umbrellalike flower cluster, adorns the tip of the plant stem. Globular, corky-ribbed fruit may be formed.

The water hemlock is related, not to the true hemlock, but to the poison hemlock, which was used to put Socrates to death. The toxic properties of the water hemlock, found throughout the plant, are concentrated in the roots. When cut, the fleshy, tuberlike root exudes an aromatic yellow oil containing the substance known as cicutoxin. Cicutoxin acts rapidly on the nervous system, causing convulsions, paralysis, respiratory failure, and death.

Also known as spotted cowbane, this poisonous plant poses a more serious problem for grazing cattle in early spring and late fall. When green forage is scarce, the normally untouched plant proves more enticing. Water hemlock then becomes a food source and is pulled up and eaten, roots included, resulting in many fatalities.

Jack-in-the-Pulpit

Arisaema triphyllum

DESCRIPTION

Classification: Wildflower
Size: Height to 3′
Characteristics: One or two three-parted leaves; spadix forms a red fruit cluster; flowers April–June

HABITAT

Swamps

RANGE

Eastern half of the U.S.A.

THE shade-loving jack-in-the-pulpit is one of the most abundant wildflowers in its range. Appearing quite early each spring, it graces the wetlands with its beauty through late fall. Delicate, translucent leaves, held like parasols over the flowering portion of the plant, may occur singly or in pairs.

Blossoming between April and June, a spike bearing tiny flowers emerges from within a striped leaf called a spathe. As this modified leaf curls over the spike like a canopy, it evokes the image of a preacher in a pulpit. To the casual observer this green spike is known as the "Jack." However, closer scrutiny of a population of *A. triphyllum* will reveal an equal number of "Jills" in the pulpit. Male and female flowers are commonly borne on separate plants. It is generally assumed that nature thus encourages cross-fertilization and a varied gene pool.

The tiny gnats of summer are largely responsible for the pollination of the spadix. The developing seeds are embedded within bright, red, fleshy berries. This fruit cluster ripens in late summer to be eaten by small animals, turtles, and a variety of birds.

A bulblike structure called a corm forms just beneath the soil. It has been reported that a freshly ground corm mixed with berries from the fruit cluster is an effective natural insecticide. The corm was a common food item for certain Indian groups that had the tenacity to prepare it. This "Indian turnip" contains sharp crystals of calcium oxalate, which can cause a severe burning sensation in the mouth. Thus, proper preparation is essential before its nutritional value may be appreciated. This is a lengthy process involving first boiling and drying the corms. After pounding and grinding, the resulting flour is then reheated and allowed to stand before being eaten.

Sometimes the jack-in-the-pulpit is called the memory plant. Anyone attempting to bite the raw turnip does not easily forget its blistering impression.

Jewelweed
Impatiens capensis

DESCRIPTION

Classification: Wildflower

Size: Height to 5′

Characteristics: Translucent pale green stems; egg-shaped serrated leaves; hanging orange-yellow flowers, spotted with reddish brown; flowers July–September

HABITAT

Swamps, marshes

RANGE

Maine to Florida; west to Texas; north to Missouri

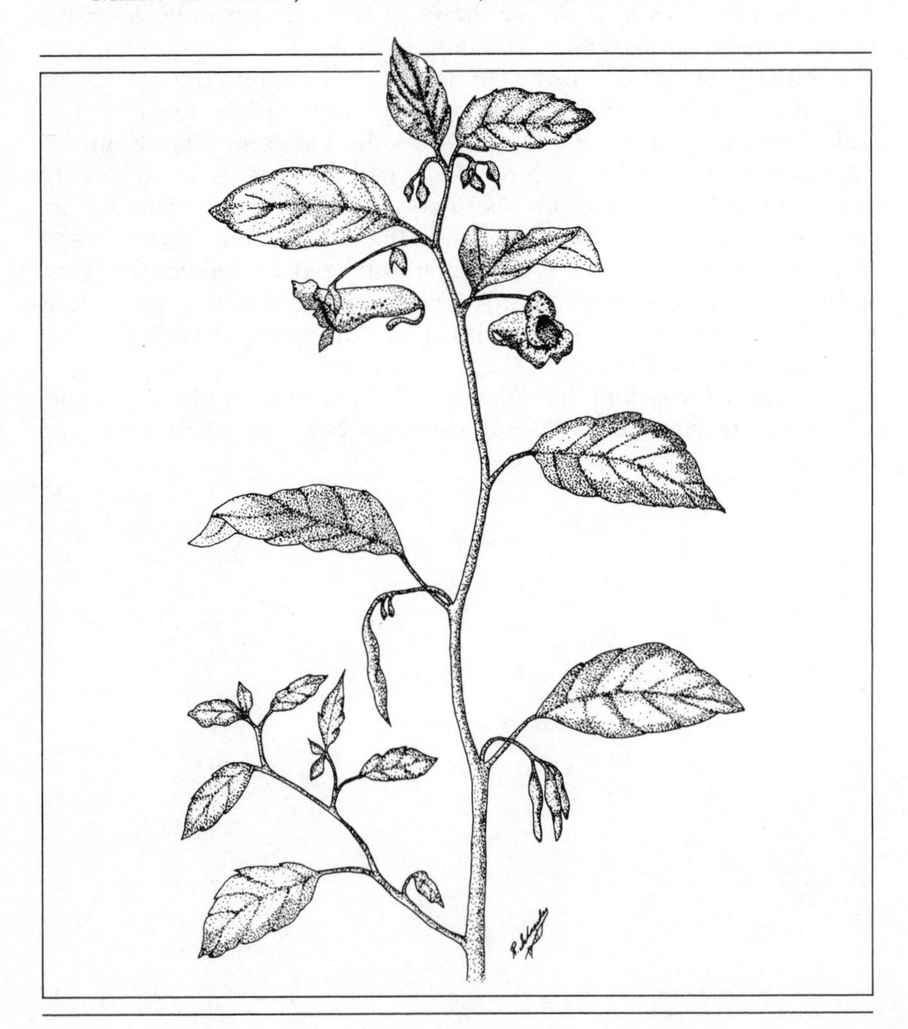

RICHLY pendent with brilliant orange-yellow, funnel-shaped flowers, jewelweed is a familiar friend in wet places. With its many fragile, branching, light green translucent stems, it is a refreshing discovery on a dewy morning. Draped heavily in round dew drops and shimmering with a silvery look, this tender, succulent annual appears to be adorned with shining jewels.

Jewelweed is also known as the touch-me-not. As the ripened seed pods are touched, they will suddenly split with the two sides curling back into tight spirals. The seeds will then be snapped out of the pod, usually with an audible pop, and be propelled several feet away from the plant, thus ensuring seed dispersion.

The jewelweed flowers can reach a length of one inch and must be forced open by bees, butterflies, and hummingbirds that seek out the deeply stored nectar, in order to ensure cross-pollination. The jewelweed also develops cleistogamous blossoms that never open but ripen self-fertilized seeds. Although these flowers will ensure plant perpetuation, inbreeding may result in an inferior plant over a period of time. Perhaps these cleistogamous blossoms are a kind of insurance against the possibility of the failure of cross-pollination.

The early settlers learned from the Indians to gather jewelweed, boil it, and use the extract. Not only was the jewelweed used to treat poison ivy, but it was also used on itchy scalps, athlete's foot, and other fungal dermatitis conditions. Scientific data has since shown that jewelweed does contain a fungicide, and it is now used in some poison ivy preparations.

Young jewelweed stems can be cooked and eaten. The seeds are also eaten raw or roasted and are said to taste like butternuts.

Showy Lady's Slipper

Cypripedium reginae

DESCRIPTION

Classification: Wildflower
Size: Height to 2½′
Characteristics: Downy stem with leaves alternating to the top; white flower with rose-mouthed pouch; flowers May–August

HABITAT

Swamps, marshes, bogs

RANGE

Maine to Georgia; west to Minnesota

DWELLING in the quiet shadows of its native haunts, the showy lady's slipper graces the area with its seductive beauty and delicate scent. Infused with a delightful free spirit, this flower gladdens the hearts of those who chance upon it. Collected and cultivated in gardens, however, the showy lady's slipper loses much of its charm, as though resigning itself to its captive fate.

The showy lady's slipper is a perennial herb with a stout, often twisted, hairy stem. Crowned with one to three striking flowers, the stem has large, prominently veined, alternate leaves along its entire length. The vivid, white, waxy side petals and sepals offer a marked contrast to the rose-colored, inflated, pouchlike lip. Resembling a dainty slipper, this pouchlike lip is often veined with deep pink or purple and contains many shallow, vertical furrows.

Well suited to cross-pollination, the flower initially lures a variety of insects with its tempting aroma. Upon seeking the nectar, a curious bee must first brush the stigma with its body, thus depositing any pollen it has brought with it. The satiated bee, in order to exit, must rub against the bursting anthers, thus ensuring the transfer of the precious pollen.

This enchanting flower harbors a dark secret. The glandular hairs on the stem and leaves secrete a skin irritant, and the handling of this plant may result in severe dermatitis. A blistering rash, similar to that produced by poison ivy, may appear within eight to twelve hours.

The showy lady's slipper, a victim of overcollection and habitat destruction, is fast disappearing. Some protection will be necessary to secure its place in our future.

Purple Loosestrife

Lythrum salicaria

DESCRIPTION

Classification: Wildflower

Size: Height to 4′

Characteristics: Spikes of vivid purple flowers held on erect stems; lance-shaped, opposite leaves, the lower ones clasping the stem; flowers June–September

HABITAT

Swamps, marshes

RANGE

Maine to North Carolina; west through West Virginia; Ohio; Indiana; Minnesota

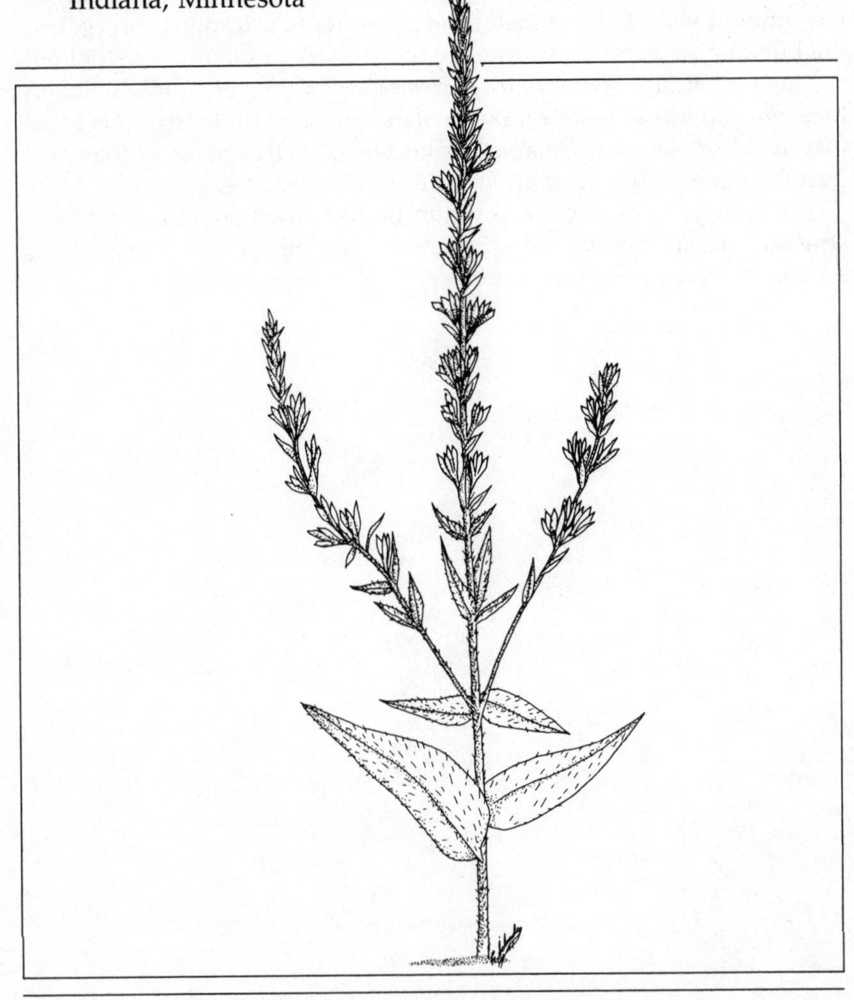

A EUROPEAN introduction, the purple loosestrife carpets acres of land with its vivid magenta blossoms. An inland marsh aglow with this showy plant provides a magnificent display. An aggressive perennial, it spreads quickly through the wetlands, crowding out native vegetation. A plant of wild beauty with its colorful flowers, the purple loosestrife can attain heights of four feet.

The six-petaled flowers are of interest to botanists because they exhibit trimorphism. Referring to three forms, trimorphism is observed in the stamens and pistils, which are of three different lengths in each blossom. The pollen from any given set of stamens is perfectly suited to fertilizing a pistil of corresponding length.

The lance-shaped leaves grow in opposite pairs or whorls of three. The lower leaves are somewhat more downy than the upper ones and grow tightly grasping the stem.

The name of the plant was derived from the early practice of placing the purple loosestrife over the yoke of quarrelsome oxen. The plant was said to help the oxen "lose their strife" and calm down.

The purple loosestrife provides cover for muskrat and waterfowl but is of little nutritional value compared to native vegetation. The plant is often considered a pest because of its excessive proliferation in an area.

The leaves were used by the American Indians to treat chronic diarrhea. If food became scarce, however, the purple loosestrife could be used as an emergency ration.

Marsh Marigold

Caltha palustris

DESCRIPTION

Classification: Wildflower
Size: Height to 2′
Characteristics: Branched, dark green stem; broad, heart-shaped leaves; golden blossoms; flowers April–June

HABITAT

Swamps

RANGE

Maine to South Carolina; west to Nebraska

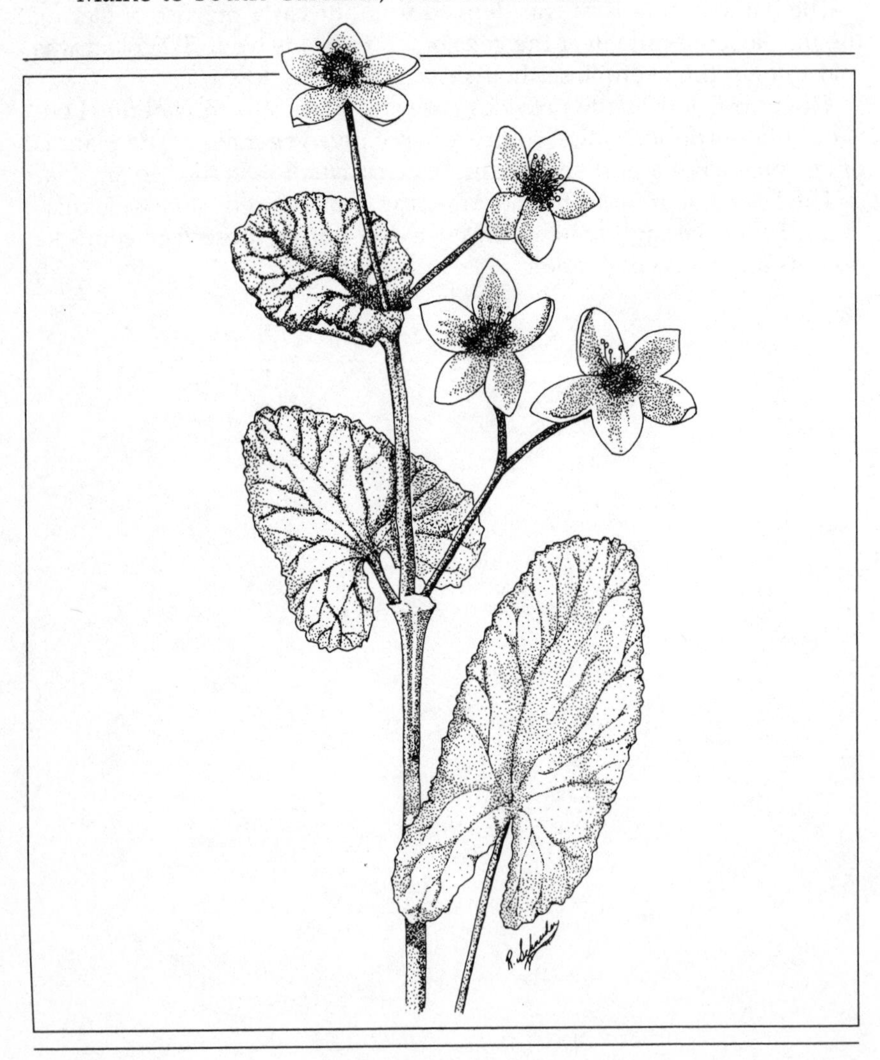

THE muted, soft green landscape of an early spring day in the swamp is annually startled by a flood of bright, golden blossoms. This flaming deluge of marsh marigolds creates such a lavish display as to be exciting to the eye and uplifting to the soul. As in eons before, this scene reenacts the ebb and flow of life and entices the mind with its infinite potential.

The marsh marigold gets its name from the Latin *caltha* meaning marigold and *palustris* meaning swamp. In reality, however, it is not a marigold but a buttercup. The bright yellow blooms, shaped like shallow cups, are not true petals but colored sepals. The pistils are pollinated by bees and *Syrphus* flies, which delight in the splendor of these sepals as much as they do any petals.

A native of North America, the marsh marigold is a common perennial. Often called a cowslip, it grows well in wet meadows where it may be eaten by an errant cow or other animal. The plant contains a powerful toxin which in small doses causes intense intestinal distress. In larger amounts the toxin is a cardiac and respiratory depressant.

In colonial times a strong brew was made from the leaves of the plant to alleviate the symptoms of coughs and respiratory infections. The toxic properties of the marsh marigold also made it effective in subduing epileptic fits. Further medicinal use includes the treatment of warts. The reported success may be the result of the high vitamin A content.

The plant can be eaten, in spite of its toxicity, when properly prepared. The harmful glycoside is destroyed by drying and cooking. The leaves and stems may be cut, boiled, and drained three or four times. This process renders the greens edible and delicious when chopped and seasoned with butter and salt.

Mint
Mentha sp.

DESCRIPTION

Classification: Wildflower
Size: Height to 2′
Characteristics: Oblong, aromatic, opposite leaves; lavender flowers; square stem; flowers July–September

HABITAT

Swamps, marshes

RANGE

Continental U.S.A. north of the Carolinas

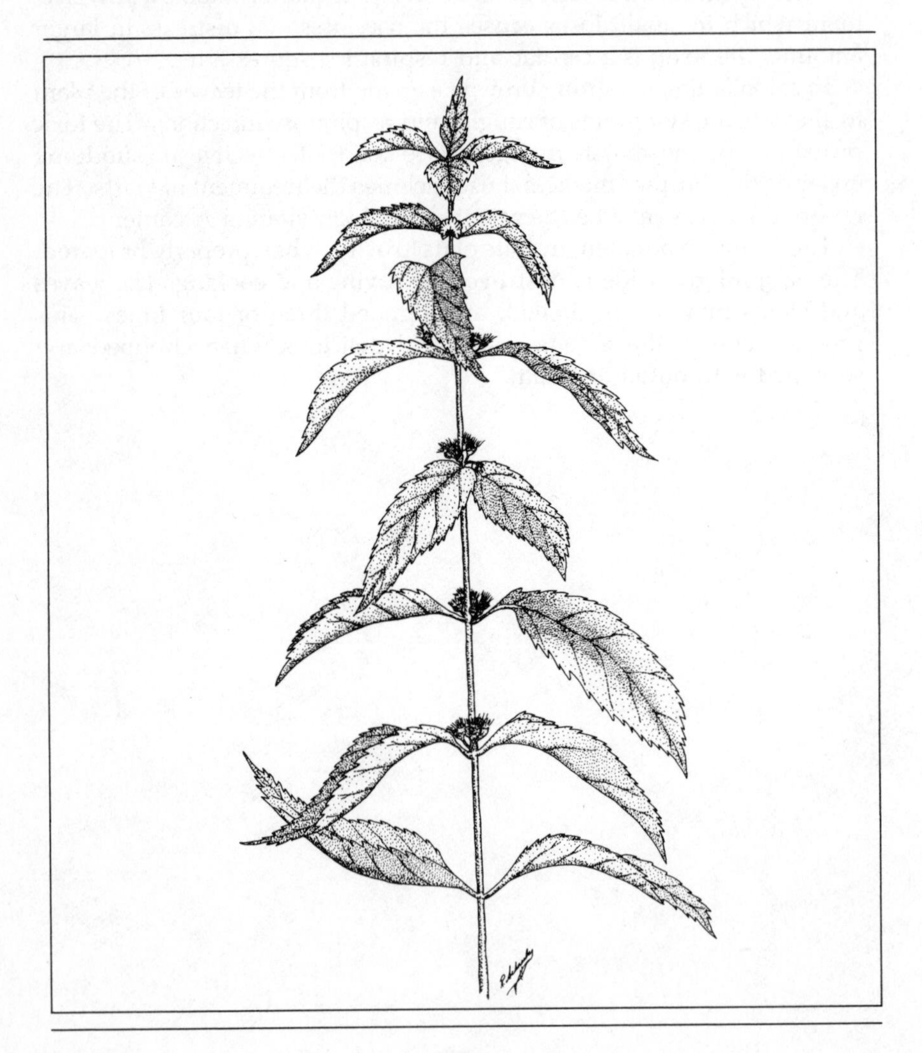

IF a marsh visitor pauses to examine more closely a pale, lavender blossom, a faint, yet pleasantly familiar fragrance may greet the senses. As a leaf is crushed, the volatile oils dissipate quickly into the surrounding area. The aroma intensifies and the unmistakable reference to mint is established.

The wild mint is the only native mint of the United States and is found mainly in the northern portion of the country. The tiny, bell-shaped purple flowers are arranged in circles around the stem of the plant and are located above each pair of opposite leaves. The sturdy, square stem, covered with fine downy hairs, supports the highly aromatic leaves.

There are many other *Mentha* species, such as spearmint, water mint, and peppermint, found growing in wet places throughout most of the United States. They commonly display purple flowers in spikes or whorls forming interrupted spikes.

Mint has been known since ancient days for its multitude of uses in medicine, scents, flavorings, and foods. Rich in vitamins A and C, the plant used to be eaten to prevent scurvy. The leaves can be boiled in water to produce a mentholated steam for a sickroom, or steeped for a day to produce a mint beverage. When dried the leaves may be used to make teas or to flavor jellies, sauces, dressing, and drinks.

Pickerelweed

Pontederia cordata

DESCRIPTION

Classification: Wildflower
Size: Height to 4′
Characteristics: Heart-shaped leaves; violet-blue flowers borne on spike; flowers June–October

HABITAT

Swamps, marshes

RANGE

Eastern half of the U.S.A.

RISING stately above the still shallows of the marsh, the pickerelweed beckons the curious naturalist. The triangular leaves form a vibrant green backdrop for the stout stem, which carries a spike of blue flowers. The common name of the pickerelweed was established when an enthusiastic observer noticed the frequency of pickerel inhabiting the same waters.

As graceful and striking a plant as it is, the pickerelweed is considered a nuisance in a water supply. Spreading rapidly, the pickerelweed clogs reservoirs and dramatically lowers the water level through excessive moisture loss due to transpiration.

The genus name, *Pontederia,* is a tribute to the great Italian botanist Giulio Pontedera. While a botany professor at the University of Padua, he first observed this plant in 1730. The species name, *cordata,* means "heart-shaped," referring, of course, to the leaves.

The spike of violet flowers is unique because each flower blooms for one day and then dies. It is followed in rapid succession by the blooming of other flowers, so that the pickerelweed is in blossom throughout the summer and into the fall. Each flower is funnel-shaped with a three-lobed upper lip. The middle lobe is marked with two yellow spots. The stamens and pistils are of differing lengths to ensure the insect contact that is needed for cross-pollination. The fruit that is formed is bladderlike and contains a single seed.

The seeds of the pickerelweed serve as food for ducks and muskrats, while the leaves of the plant are food for deer. The Indians also used the young leaves in salads and ate the seeds like nuts. When roasted, the seeds were added to cereals and breads or ground and made into flour.

Pitcher Plant

Sarracenia purpurea

DESCRIPTION

Classification: Wildflower

Size: Height to 2′

Characteristics: Hollow leaves shaped like a pitcher; leaves may be richly striped in yellow, green, and red; flowers May–July

HABITAT

Swamps, bogs

RANGE

Maine to Florida; west to Minnesota

OFFERING sweet nectar to a kaleidoscope of insects, an innocuous-looking glade of goblets lures the unwary victims to their doom. The leaves of the pitcher plant secrete an insect-attracting syrup and are bordered at the mouth by downward-pointing bristles. Held within the vessellike leaves, this liquid often contains the remains of drowned insects. It appears that the insects are part of the plant's nutrition, but this deduction has been seriously questioned. There is at least one mosquito, *Wyeomyia smithii*, whose larval stage is completed in this predator-free environment. The adult female lays her eggs on the newly opened leaves of the pitcher plant. As water gathers within the vessel, the eggs begin to hatch. The larvae remain there throughout the winter, frozen in the cores of ice, to emerge as adults the following spring. *Wyeomyia smithii* is a vegetarian and is harmless to animals.

There are seven distinct species of pitcher plants found along the east coast. They are prominent in very acidic northern peat bogs, where they root in the vegetative mats rather than directly in the soil. In the central and southern regions the pitcher plants are found in acid swamps where low nitrate levels prevail. Hooded like a cobra about to strike, the western pitcher, *Darlingtonia californica*, is a bit more slender than its east coast cousins. Throughout the United States, the curious leathery leaves of the pitcher plants are evergreen, remaining through the year despite harsh weather conditions.

Globular, nodding flowers appearing in the late spring are held on a long stalk as much as two feet above the pitchers. These flowers are usually deep purple or yellow and may be more than two inches in diameter. Since the flower resembles a pocket watch dial without hands and is therefore mute, the plant is often called a "dumb watch."

Also referred to as Indian cup, the pitcher plant was long regarded by Native Americans as an aid to decreasing the term of smallpox and preventing the formation of deep pits during convalescence.

With destruction of habitat, the pitcher plant has become seriously threatened in certain areas. Given reasonable care, however, it is a hardy plant, and populations can generally be reestablished.

Rose Pogonia

Pogonia ophioglossoides

DESCRIPTION

Classification: Wildflower

Size: Height to 2′

Characteristics: Single leaf found near middle of slender green stem; single, ¾″, rose pink flower with bearded and fringed lip located at tip of stem; flowers May–August

HABITAT

Swamps, marshes, bogs

RANGE

Maine to Florida; west to Texas, Illinois, Indiana, and Missouri

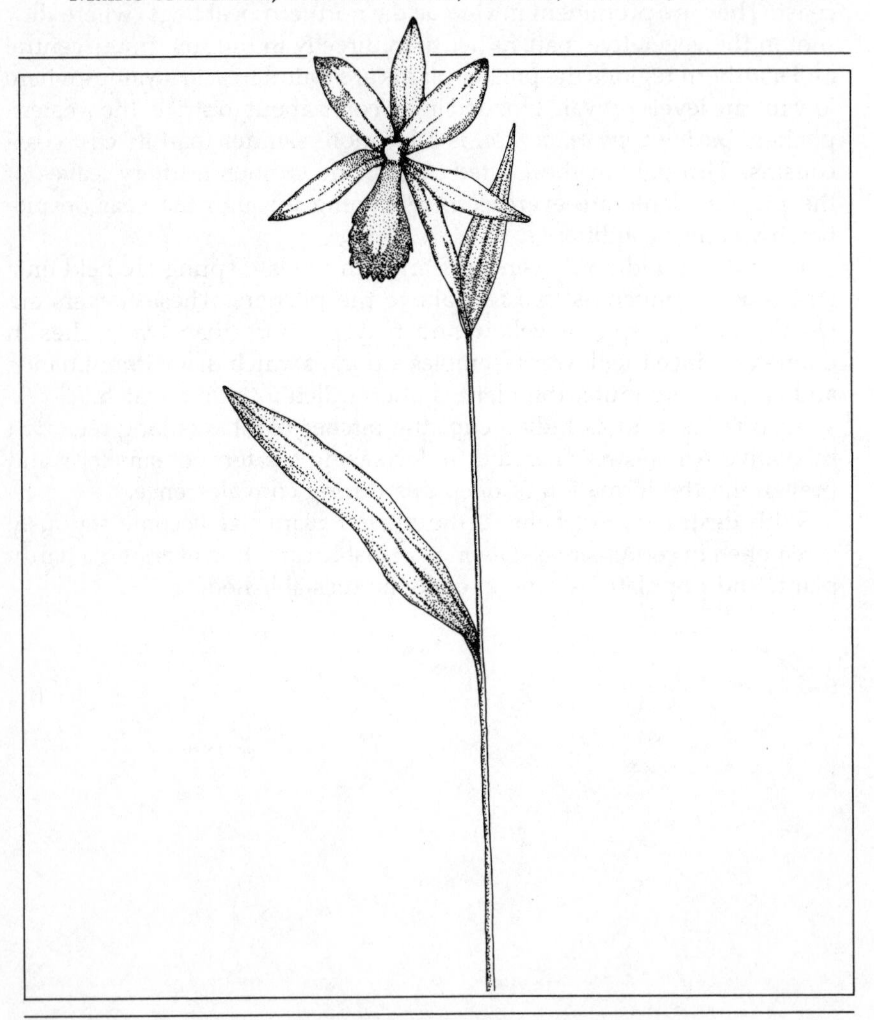

IN early summer, the rose pogonia can be found growing in the company of the glistening sundew, the sturdy sedge, and the soft sphagnum moss. The delicate, rose pink shades of its fragile flower lend a soft blush to the young green growth of the bog, while its violetlike fragrance perfumes the air.

The rose pogonia is found in places in the eastern United States where soil condition, rather than temperature, is the controlling factor. Each pink flower, an isolated orchid, rises on its own stalk. The warm pink petals and sepals offer a muted background for the vibrant yellow bearded lip. A single, lance-shaped leaf clasps the middle of the slender, green stem while a leaflike bract grows below the single flower. The long-lasting flower of the rose pogonia is said to resemble a snake's mouth because of its contrasting coloration, hence the species name *ophioglossoides*, "ophio" meaning snake and "glossoides" meaning mouth.

The rose pogonia flower is designed to protect the plant from self-fertilization. Upon entering the flower chamber for the sweet nectar, the insect, usually a bee, is pressed tightly by the flower and must back out in order to leave the chamber. In doing so the insect catches pollen on its head and thus carries it to other plants.

The aromatic pod, from which vanilla is derived, comes from the fruit of an orchid belonging to this group. Since it is very difficult to duplicate the chemical structure, natural vanilla is still highly valuable, and the orchids are carefully pollinated and cultivated.

Skunk Cabbage

Symplocarpus foetidus

DESCRIPTION

Classification: Wildflower
Size: Height to 3′
Characteristics: Large, deep green, cabbagelike leaves with elongated stems; mottled, purplish green hoods; flowers February–April

HABITAT

Swamps, marshes

RANGE

Maine to Georgia; west to Missouri

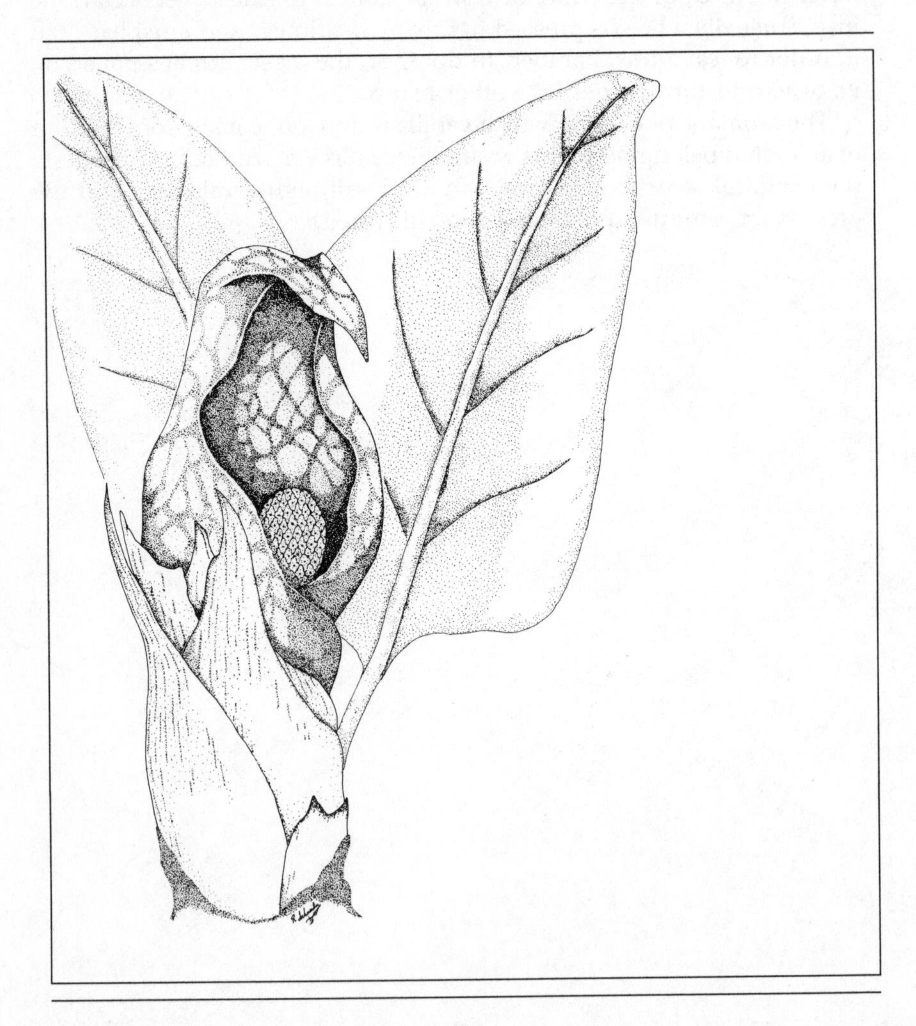

THE skunk cabbage is truly the harbinger of spring as it eagerly thrusts its way through the swamp floor on a bright, balmy winter day. Since the plant's development begins in the fall, the skunk cabbage is ready to open at the first fleeting hint of warmth.

Venturing into the quiet, leafless swamp in winter, few people are aware of the activity already underway in the skunk cabbage. Although the purple and green hoods of the plant lie frozen in the glistening ice, they are busily engaged in cellular respiration. The heat generated by this process can actually raise the temperature of the plant more than twenty degrees above the surrounding air and melt any snow in the immediate vicinity. As the tightly closed hood, or spathe, uncurls, the pungent odor of the plant permeates the clean, fresh smell of the crisp, winter air. Aptly named because of its unpleasant scent and its large, cabbagelike leaves, the skunk cabbage begins to enliven the dreary landscape, first with its multicolored spathes, then with its bright green foliage.

Borne on a thick fingerlike stalk called a spadix, the tiny, delicate, flesh-colored flowers are encased by the spathe. This arrangement offers them excellent protection against the brazen cold days that heckle the spring thaw.

The skunk cabbage is pollinated by small carrion flies, which are attracted by its odor of decaying flesh. Successful pollination results in the formation of a pulpy mass of scarlet berries.

After flowering, the large leaves begin to appear and spread rapidly throughout the area. Unfolding to lengths of two feet, the leaves eventually develop elongated stems. Thick rootstocks form, and the fleshy, fibrous roots bind the soil, enhancing it for later spring growth.

Skunk cabbage is rich in caustic calcium oxalate crystals, which can cause intense burning in the mouth, vomiting, and temporary blindness. Even extensive processing may not completely eliminate these hazards. It is probably best to avoid ingesting this plant.

Sundew

Drosera rotundifolia

DESCRIPTION

Classification: Wildflower
Size: Height to 10″
Characteristics: Basal rosette of round reddish green leaves covered with sticky hairs; white flowers held on single erect stalk; flowers June–August

HABITAT

Bogs

RANGE

Maine to Florida; west to Texas

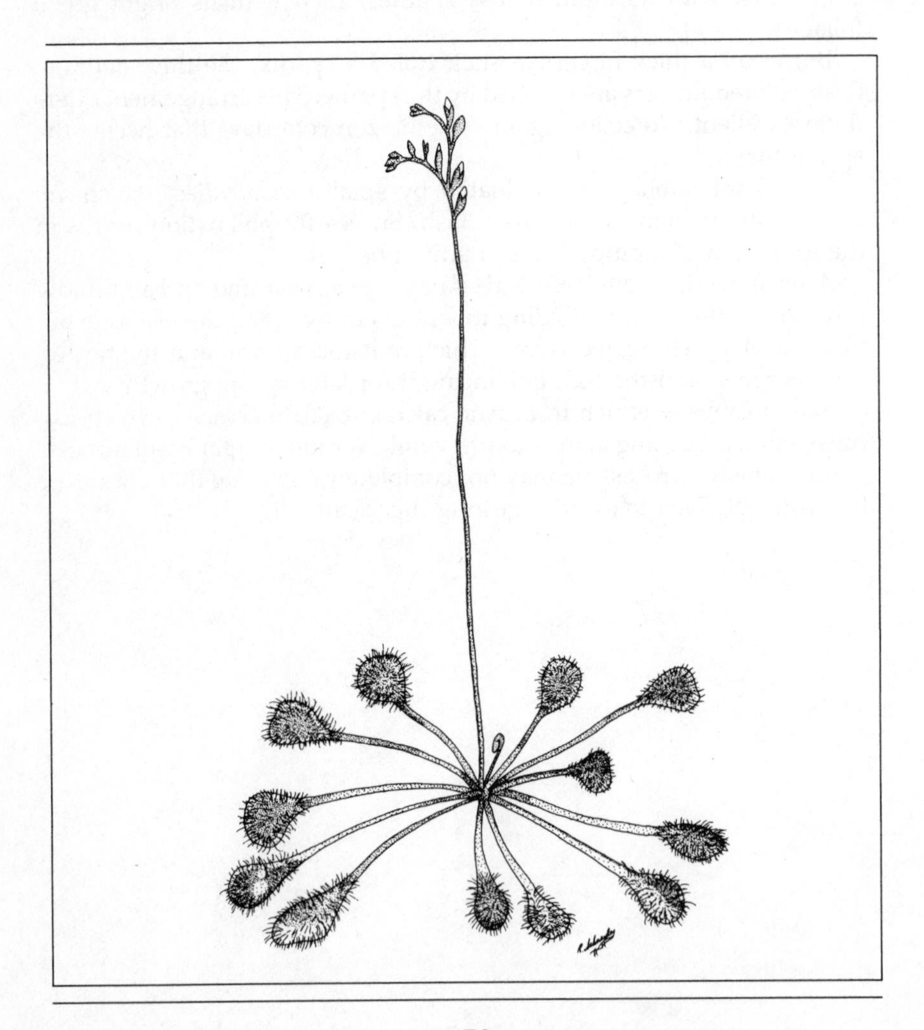

SPARKLING in the sunlight, the sundew tantalizingly beckons to insects one and all. Covered with numerous tiny tentacles that exude droplets of a powerful adhesive, the reddish green leaves appear to be drenched with dew. Growing in nitrogen-deficient, acid-soil bogs, the small and inconspicuous sundews are deadly insect traps which seldom release their victims alive.

Attracted by the coloration and the sweet secretions of the sundew, the insect is drawn to the plant. Landing on even a single tentacle, it is held fast. As the insect struggles in vain, impulses from one tentacle spread rapidly to others. In a choreographed wave the tentacles bend over and smother the prey. A sundew can kill its victim in less than fifteen minutes, although digestion may not be complete for several weeks. The insect's body is made soluble by enzymes and is absorbed by the tentacles. Afterward, the tentacles relax and resume their sinister pursuits. A light breeze can dislodge the empty hulk of the insect and make room for a new catch.

Since sundews grow in areas where mosquito larvae are abundant, adult mosquitoes are a large part of their diet. Seeking a perch to dry themselves, the newly hatched adults naively climb onto the sundew plant and meet their doom.

In midsummer a slender flower stalk emerges to tower above the basal plant. Delicate white to red blossoms open one at a time and attract tiny flies and wasps. Avoiding the treacherous leaves, these insects gather the pollen oblivious to the danger. Upon pollination a many-seeded capsule is formed.

While the carnivorous sundew enjoys a varied insect diet, its deadly secretions are innocuous to the assassin bug. Spending its life among the tentacles and mimicking the color of the plant, the assassin bug boldly seeks out the helplessly ensnared victims. Upon locating a hapless prey, this opportunist first injects the insect with a potent neurotoxin. Using its powerful proboscis, the assassin bug then sucks out the victim's body contents for its own nourishment.

There are many species of sundews. Two very common groups include the spatulate-leaved sundew, *Drosera intermedia*, and the thread-leaved sundew, *Drosera filiformis*, both named for the shape of their leaves.

Sundews were once collected and brought indoors to serve as flypaper. Early settlers also used them to extract a red fluid to be used as ink.

Turtlehead

Chelone glabra

DESCRIPTION

Classification: Wildflower
Size: Height to 3′
Characteristics: White flower blossoms resembling the head of a turtle; opposite, serrated, lance-shaped leaves; flowers July–September

HABITAT

Swamps, marshes

RANGE

Maine to Georgia; west to Kansas; north to Minnesota

ON a sunny day in late summer, the creamy white flowers of the turtlehead may first be seen. Flourishing luxuriantly in marshes, the plant is more prominent and peculiar than pretty. The flower of the turtlehead, as viewed from the side, resembles the head of a turtle or snake with its mouth open.

Located at the tip of the slender, erect stem, the flowers are arranged in a blunt spike. The flower buds may be numerous, but they do not all open at the same time. Blooming at intervals, beginning with the bottom blossoms and proceeding upwards, these buds burst into white or faintly tinged pink two-lipped flowers.

The long, narrow leaves of the plant grow opposite and are entirely toothed. Due to their unusually short petioles, the leaves appear to be growing directly out of the stem.

The checkerspot butterfly is often found in the vicinity of the turtlehead, since the leaves of this plant are food for its caterpillar stage. The caterpillars feed until autumn, hibernate for the winter, continue feeding briefly in the spring, and then pupate. Upon emerging, the butterflies mate and lay new eggs on the leaves, thus perpetuating the cycle.

The turtlehead blooms from late summer until the first hard frost. It is often confused with the foxglove beard-tongue, *Penstemon digitalis*, by beginners. An important difference to be noted is that the turtlehead flowers sit on a spike, while the foxglove flowers form a loose spray.

Venus Flytrap

Dionaea muscipula

DESCRIPTION

Classification: Wildflower
Size: Height to 1′
Characteristics: Basal leaves with lengthwise folded blades; white flower cluster on a leafless stalk; flowers May–June

HABITAT

Swamps, marshes, bogs

RANGE

North and South Carolina

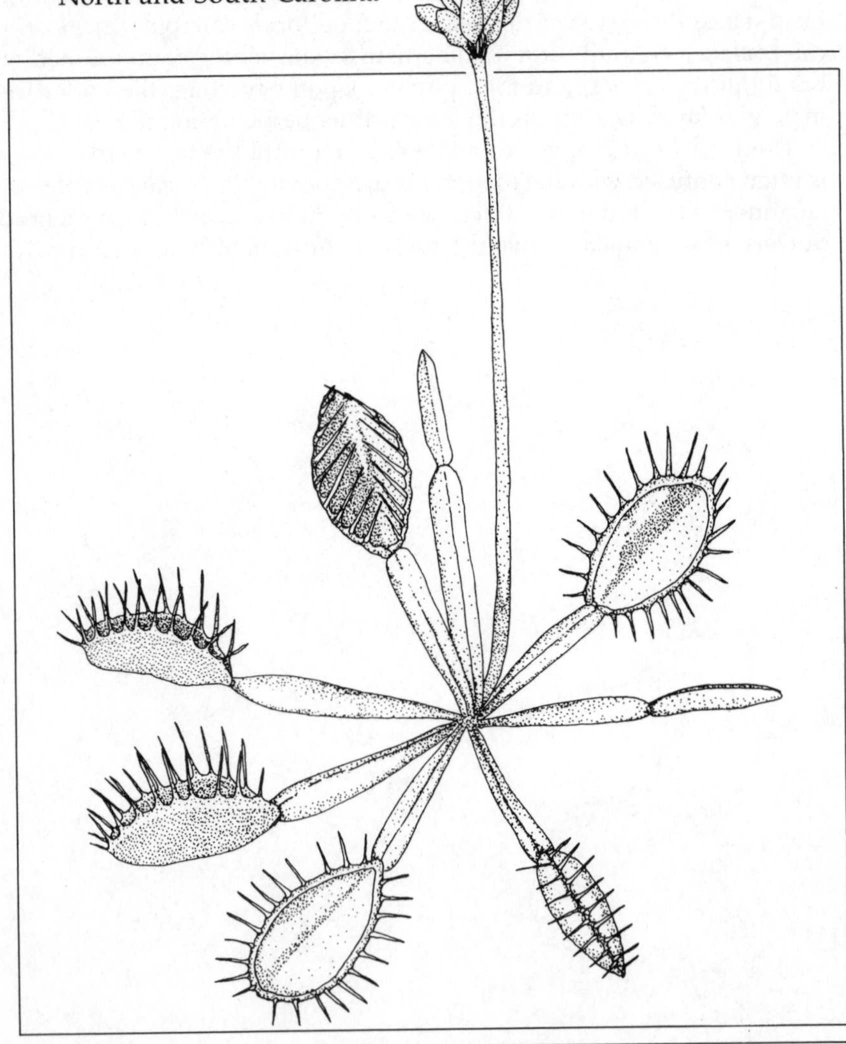

EXUDING a sweet nectar, the leaves of the Venus flytrap gently entice a nearby insect. As the insect alights, it brushes against the trigger hairs. The hinged leaf quickly snaps shut, trapping the victim and forming an impenetrable prison with its long, stout, interlocking spines. The insect is then devoured.

The Venus flytrap is a mechanical marvel, and for its small size probably the most effective prey-trapping land plant. The hinged leaves, fringed with long, stiff bristles, contain three trigger hairs on each side of the midrib. When these hairs are touched twice, the leaf closes like two upper eyelids coming together and forms a cage where the intruder is held and digested. This capturing movement takes only one or two seconds, and if the plant reacts to an errant twig or pebble, the leaves are reset.

After capture, the trap remains closed for several days. During that time it secretes a digestive enzyme that breaks down the animal protein into amino acids and peptones. These substances can then be absorbed and used by the Venus flytrap. Growing in nitrogen-deficient soil, the plant has evolved a carnivorous lifestyle in order to meet its nutritional needs.

The Venus flytrap is an attractive plant with its showy cluster of five-petaled, white flowers borne at the tip of its leafless stalk. Its basal hinged leaves are green on the outside and orange-pink on the inside.

In spite of a restricted natural range, *D. muscipula* has become a very popular house plant due to its unique botanical properties. It has also served as a model for many science fiction horror stories because its movements simulate aggression. The Venus flytrap is classified as an endangered species in both North and South Carolina.

Water Plantain

Alisma subcordatum

DESCRIPTION

Classification: Wildflower
Size: Height to 3′
Characteristics: Small, white flowers (½″) arranged in whorls on ends of stiff, branching structures; olive green, elliptical, basal leaves with long petioles; flowers June–October

HABITAT

Swamps, marshes

RANGE

Maine to Florida; west to Wisconsin

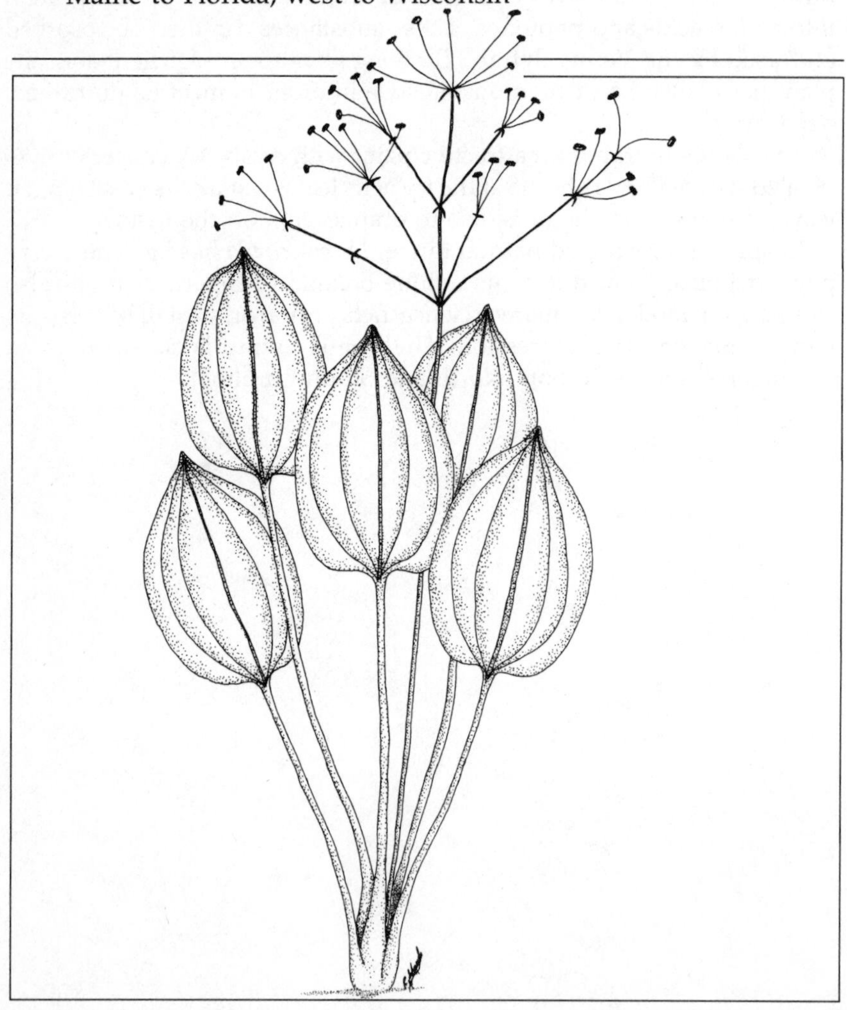

THE distinctly veined, elliptical emergent leaves of the water plantain rise ostentatiously above the shallow water of the marsh. Forming a thick, olive green canopy over the water, this plant thrives along wetlands margins. In contrast to the oval shape of the emergent leaves, the submerged leaves of the water plantain exhibit a ribbonlike shape. This adaptation allows water currents to sweep easily through the vegetation without uprooting the plant.

The water plantain is a perennial with a stout underground stem. It is capable of surviving in both drought and flood conditions. However, in order to produce new plants each spring, the rootstock must be submerged.

Small white flowers are borne in whorls on many tall, spindly branches arising from the leaves of the plant. These miniature three-petaled, three-sepaled flowers usually go unnoticed except by the insects that pollinate this plant. After pollination, nutlets are produced and crowded into fruits with two ridges. These nutlets have been reported as constituting a small part of the diet of wild ducks and pheasants but are not considered an important food item for most game species. The rootlike structure is edible to some animals. When handled by humans, however, it may cause a mild skin irritation.

Speckled Alder

Alnus rugosa

DESCRIPTION

Classification: Shrub
Size: Height to 30′
Characteristics: Deciduous; smooth bark; alternate oval leaves; several trunks; blossoms produced in catkins; flowers March–May

HABITAT

Swamps, marshes

RANGE

Maine to West Virginia; west to Iowa; north to North Dakota

AMONG the winter silhouettes of the dogwoods, willows, and red maples of the swamp, a densely tangled shrub bearing small blackened cones stands virtually unnoticed. Thriving in nitrogen-poor soil, the speckled alder can grow anywhere but prefers the moist soil of the wetlands. As spring warms the swamp, the alder blossoms into full glory as it displays a splendid crown of innumerable purple and yellow catkins. During the other seasons of the year, the speckled alder is often overlooked because it has no colorful flowers or fruits.

The bark of this shrub is a cinnamon brown color richly speckled with white lenticels, hence its common name. The alternate, oval leaves have toothed edges and may be fuzzy along the veins on the undersides.

In the spring before the leaves emerge, male and female flower clusters form separate catkins on the same plant. The dangling male catkins, which are several inches long, bloom and spread their pollen by the wind. The female catkins, which are shorter, have sticky reddish hairs projecting from them on which they catch the pollen. After fertilization, the seed, which is a small, flat nutlet, is produced within a fruiting cone.

The alder is invaluable in the wild. Not only do its roots create dense mats to hold the soil in place, but once it is established the alder spreads rapidly by vegetative growth. New shoots form from old bases, from underground stems, or by layering, a process in which a branch that comes in contact with the ground sends out roots and becomes a new plant.

The alder is host to some attractive tiny beetles, namely the alder fleabeetles, *Altica bimarginata*, which are steel blue, and the *Lina interupta*, which are patterned with vivid yellow and black. The adults of both species overwinter in the ground beneath the leaves and in the spring crawl into the alder. Feeding on the leaves, the females lay eggs, which then hatch into small dark-colored grubs. These larvae also feed on the leaves, and when they mature they attach to the undersides of the leaf.

In the summer, the shrub provides a cool, leafy cover for birds and mammals. Muskrats, moles, and shrews maintain a network of tunnels through the alder roots, while blackbirds and goldfinches build nests in the branches. The leaves, buds, and seeds of the alder provide food for ptarmigan, grouse, chickadees, and goldfinches. Deer, rabbit, beaver, moose, and muskrat browse the alder, while bees use the pollen.

From late summer to fall, large masses of aphids congregate on the stems to live off the sap. These are the alder blight aphids, *Prociphilus tessellatus*. Exuding a by-product of the sap through pores on their backs, these insects produce what looks like a mass of white cotton and are commonly called woolly aphids. Early in summer, they live on red maple leaves.

Another insect visitor, the wanderer, *Feniseca tarquinius*, a small brown and orange butterfly, may be seen around the alder. It feeds on the honeydew secretions of the aphids that have dripped on twigs, and it lays its eggs among the insects by quickly landing directly on the aphids and darting across them. The butterfly larva hatches and feeds on the bodies of the aphids, making a silk covering for itself out of the empty skins of its prey. The larva takes ten days to mature, and when fully grown it moves to pupate on a nearby leaf. In a single year the wanderer may have three broods throughout its range, the entire eastern half of the United States. The wanderer is the only carnivorous butterfly larva in this country.

Rugosa, the latin name for the alder, means "wrinkled" and refers to the leaf surfaces. The speckled alder is too small to be of much economic importance, although it does have limited uses as an ornamental and as a riverbed protector. The alder can thrive in poor soils because of the bacteria-filled nodules on its roots. These bacteria are able to use atmospheric nitrogen in the formation of protein.

During colonial times, charcoal made from alders was used to make black gunpowder. The bright orange wood is a source of dye and has been used throughout history to color animal skins as well as cloth. The wood also contains a powerful emetic, which was used by the early colonists to purge the body.

The alder has always had a special significance to the Norsemen. Legend has it that the first man was made from an ash tree while the first woman was made from an alder.

Highbush Blueberry

Vaccinium corymbosum

DESCRIPTION

Classification: Shrub
Size: Height to 15′
Characteristics: Deciduous; alternate elliptical leaves, smooth above, slightly fuzzy below; multistemmed; white or pale pink bell-like blossoms; blue berries; flowers May–June

HABITAT

Swamps, bogs

RANGE

Eastern half of the U.S.A.

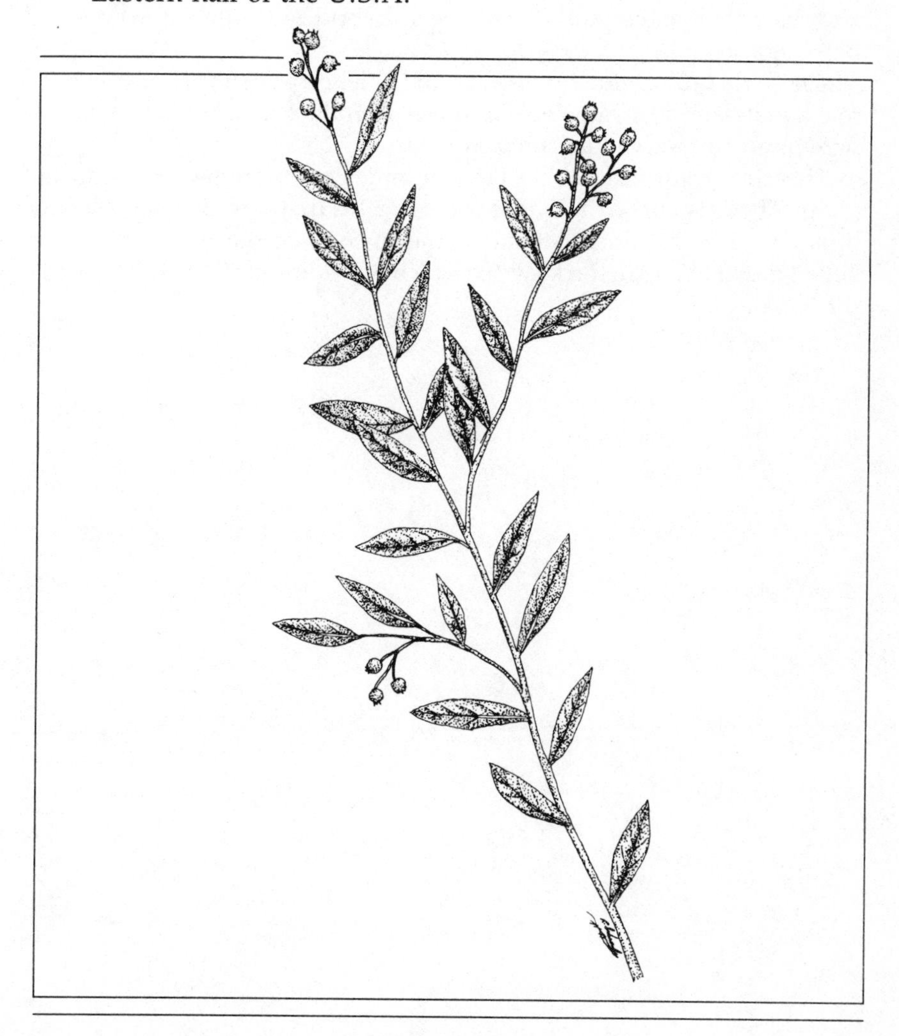

IN the autumn, the highbush blueberry is one of the most variable shrubs. Changing from green to yellow to orange to crimson, its graceful branches are vibrant and reflect intimately the spirit of that glorious season when the earth is ablaze with color.

The blueberry is a tall, multistemmed shrub, with alternate, elliptical leaves that are green both above and below. A member of the heath family, the blueberry benefits from an acid environment such as a bog, since its root fungus partners grow best in this soil type.

Leaves appear in early spring, and the white or pale pink bell-like flower clusters soon follow. Pollination is accomplished by bees, and by midsummer the plant is bursting with succulent blueberries. The fruit is an excellent food source for rabbit, grouse, pheasant, dove, muskrat, otter, mink, beaver, black bear, and naturalists who especially adore the sweet berries. The twigs of the plant are also eaten by deer and rabbits.

A closely related species of blueberry is found on the Pacific coast and many cultivated strains have been hybridized from the wild blueberry species.

Since it is propagated by suckers, the blueberry bush is relatively easy to cultivate and is a valuable commercial product. Its fruit is sold fresh, frozen, and preserved as jam and jelly.

This delightful fruit was a tasty summer treat for the American Indians. They also used the leaves to make a tea that was given as a spring tonic. To cure an upset stomach, the juice from the blueberries was boiled down to form a thick syrup, which was administered orally several times a day.

Bog Rosemary

Andromeda glaucophylla

DESCRIPTION

Classification: Shrub
Size: Height to 3′
Characteristics: Evergreen; narrow, toothless, alternate leaves with curved margins, dark green above, whitened beneath; smooth twigs; small white to pink urn-shaped blossoms; flowers May–July

HABITAT

Bogs

RANGE

Maine to West Virginia; west to Indiana and Minnesota

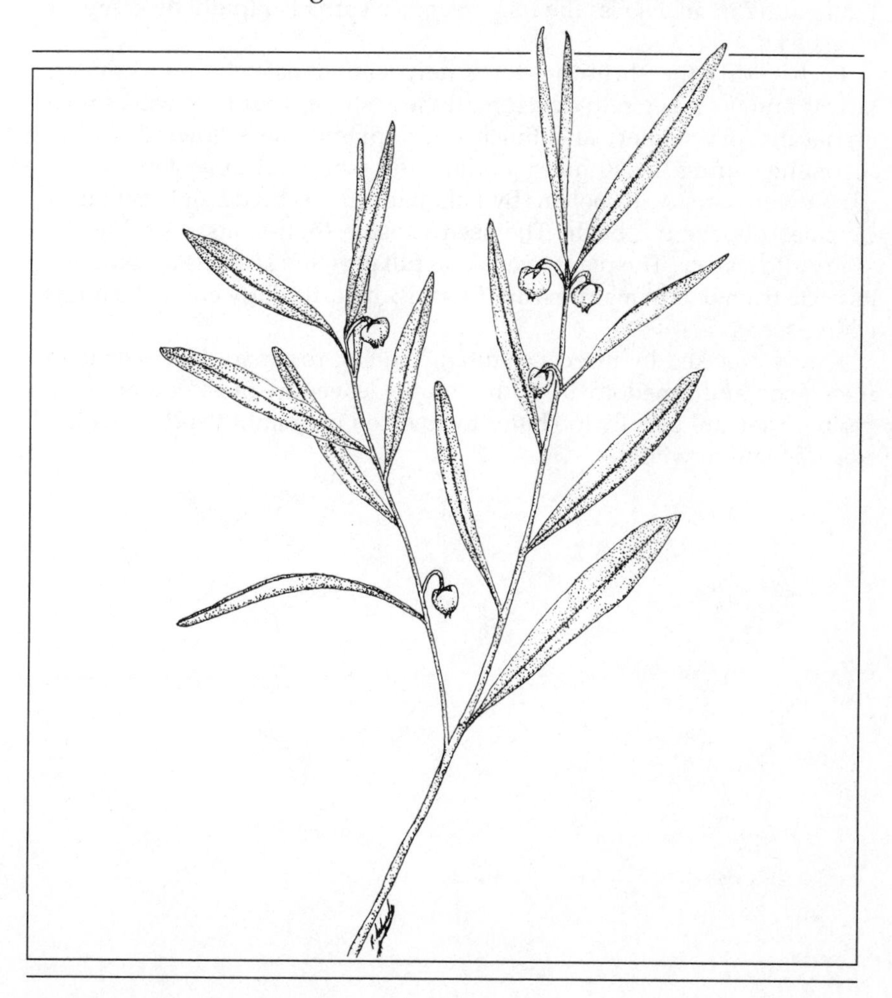

A SPARSELY branched shrub with thick, leathery leaves huddles on the edge of the sapphire blue watery eye of the bog. "This plant is always fixed," wrote Linnaeus, "in the midst of swamps, as Andromeda herself was chained to a rock in the sea which bathed her feet as the fresh water does the roots of this plant."

The bog rosemary, or *Andromeda,* is of particular interest to botanists because it is capable of withstanding dry environmental conditions although it thrives in wet areas. Speculation suggests that since the high acidity of the bog interferes with water absorption, these plants are structurally designed to reduce water loss. Thus the leathery leaves and rolled leaf margins serve the bog rosemary well in maintaining an acceptable water content.

Andromeda is a low-growing evergreen with crowded, narrow, dark green leaves. The lower surface of each leaf, veiled by tiny white hairs, distinguishes this shrub from the other members of the heath family. Flourishing in acid soils, the bog rosemary spreads rapidly by creeping rootstocks.

Each spring the shrub produces new leafy shoots. By midsummer, flower buds will begin to develop on these shoots, but they will remain immature, overwinter, and finish development the following spring. Blooming during the summer months, the white, carmine-stained bell-like flowers burst with pollen. By fall, mature, erect reddish brown fruit capsules pepper the bush. The seed capsule then splits releasing the seeds into the air. The plant needs two full growing seasons to complete its cycle from bud to mature fruit. Ptarmigan particularly enjoy the nourishing seeds.

Rarely attacked by insects or fungi, the bog rosemary leaves contain the poison andromedotoxin. Although the leaves are most potent in the spring, they are usually too bitter to be eaten in quantity and so are not troublesome to wildlife.

Buttonbush

Cephalanthus occidentalis

DESCRIPTION

Classification: Shrub
Size: Height to 30′
Characteristics: Deciduous; opposite or whorled leaves, deep green above, paler below; white blossoms in globular heads; gray-brown to black, ridged bark; flowers May–August

HABITAT

Swamps

RANGE

Eastern half of the U.S.A.; west to New Mexico, Arizona, and California

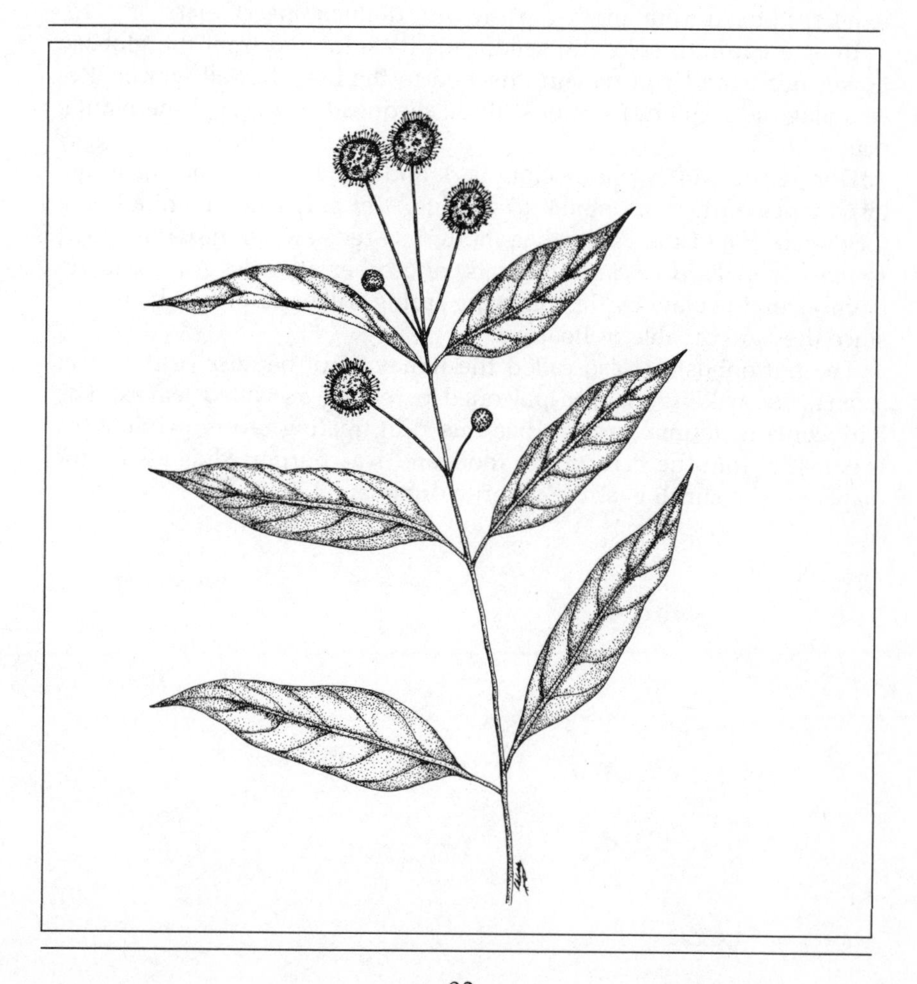

ON a warm, summer evening, a sweet fragrance gently wafts through the swamp. Emanating from the aquatic buttonbush, this delightful aroma is released by the globular flower heads. Resembling pins stuck in a pincushion, each minute blossom has four creamy white petals with elongated styles.

The buttonbush is noted for its ability to withstand flood conditions and grows quite well among the alders and willows. It is a favorite nesting site for the song sparrow and the red-winged blackbird with its ascending, widely branched, reddish-tipped stems.

By late spring the buttonbush sports glossy green leaves growing opposite or in whorls of three to four, and by late summer, it is filled with globes of blossoms.

Pollination involves a "brush type" system. Surrounding an immature pistil, the anthers shed their pollen onto its tip. As the flower blooms, the pistil elongates and carries with it the pollen that has become deposited. When an insect comes along and removes the pollen, the tip of the pistil then matures and becomes sticky. Now the pistil is ready to accept pollen from another plant with the next insect visit.

By late autumn, spherical seedheads have formed from the globose flower heads and will remain attached to the branches all winter. Resembling old-fashioned buttons, these seedheads suggested the plant's common name.

During the winter, the peeling bark and scraggly branches belie the importance of the buttonbush to wildlife. Not only is the plant a home for the cocoon of the Promethea moth, its seeds are an important food source for mallard ducks, muskrats, and other animals. Any seed remaining on the plant by the following spring will be dispersed by water, since they are capable of floating.

The buttonbush is also called the honey plant because of its sweet aroma, but wildlife may be poisoned by eating its wilted leaves. The bark contains tannin and has been used in treating fevers, while a tea was made from the bark of the roots and was purported as a cure for diabetes. The shrub is also a favorite ornamental.

Labrador Tea

Ledum groenlandicum

DESCRIPTION

Classification: Shrub
Size: Height to 4′
Characteristics: Evergreen; densely hairy twigs; alternate leathery leaves with rolled edges; woolly underneath; terminal clusters of white blossoms; flowers May–July

HABITAT

Bogs

RANGE

Maine to northern New Jersey; west to Ohio; north to Minnesota

THIS small, boreal evergreen shrub brightens the springtime bog. With its umbrellalike clusters of startling white flowers set against its glossy, green leaves, the Labrador tea revives a winter-weary spirit to envision the balmy days to come.

The Labrador tea is a low, compact shrub with extremely velvety hairy twigs. Its narrow, leathery leaves are dark green above with light, rusty brown woolly undersides. Having a rolled margin and a somewhat thick surface, the leaves are constructed to prevent water loss. Even in a sodden bog, the acid water may inhibit a plant's water-absorption capability, thus making vital any adaptations for water-loss reduction.

L. groenlandicum is closely associated with the leatherleaf and bog rosemary, both of which have hairless or nearly hairless twigs. The leaves of the Labrador tea are fragrant when crushed and were used for tea during colonial times. After flowering through the spring and into the summer, the flowers that have been pollinated develop into nodding clusters of elliptical fruit capsules.

Leatherleaf

Chamaedaphne calyculata

DESCRIPTION

Classification: Shrub
Size: Height to 4′
Characteristics: Evergreen; alternate, oblong, leathery leaves, yellowish below; white bell-like blossoms; hairless twigs; flowers March–July

HABITAT

Swamps, bogs

RANGE

Maine to Georgia; west to Ohio and Wisconsin

THE icy fingers of the sphagnum moss cling to the feet of the leatherleaf. Melting slowly, the frozen earth of the bog yields almost reluctantly to spring. As the chilling days wane, the swelling leatherleaf buds become more prominent, and by late spring, the white, bell-like flowers dangle gracefully from the tips of the higher branches.

Following pollination by bees, ripened fruits form small, upturned cups called "rattleboxes." Contained within these structures are the seeds of the plant, which will be held until jostled out by wind, rain, or animals.

The delicate leatherleaf thrives in the acid peat soil of the bog along with its close neighbors, the sphagnum moss, the sedges, the cranberry, and the bog rosemary. It is especially important to bog succession in that it grows out over the sphagnum moss. In this way, it encourages other plants to gain a footing on the reinforced ground.

The genus name of the leatherleaf, *Chamaedaphne*, is derived from *chamai* meaning "on the ground," and *daphne* meaning "laurel," which refers to being a low evergreen. It is a hardy shrub whose blossom number depends on the height of the snow during the previous winter. If completely covered by snow, almost all of its flowers will bloom. If mostly covered with snow, except for the top, 30 percent of its flowers will bloom. If totally exposed to the harsh winter winds, only one or two of its flowers will bloom.

The leatherleaf grows two types of branches and two types of leaves. Each summer the first growth is the main vegetative vertical branch with large leaves. Sprouting horizontally off this is a smaller twig with flower buds and smaller leaves. These two branch forms (main vertical and horizontal sprouts) can be seen throughout the fall and winter.

In the following spring, once the flowers have bloomed and the seeds matured, the small horizontal twigs shed their leaves and die. Meanwhile, the main vertical branches keep their leaves for another year or two before finally meeting their demise. New vertical stems form annually to maintain the health and vigor of this shrub.

There appears to be some disagreement as to the use of the leaves for tea since they do contain a poison called andromedotoxin. Birds, especially grouse, enjoy the fruit throughout winter and the following spring, while rabbits nibble the entire plant year round.

Swamp Loosestrife

Decodon verticillatus

DESCRIPTION

Classification: Shrub
Size: Height to 8′
Characteristics: Deciduous; arching stems; lance-shaped opposite leaves, tufts of magenta blossoms in upper leaf axils; flowers July–August

HABITAT

Swamps, marshes, bogs

RANGE

Maine to Florida; west to Louisiana; north to Illinois

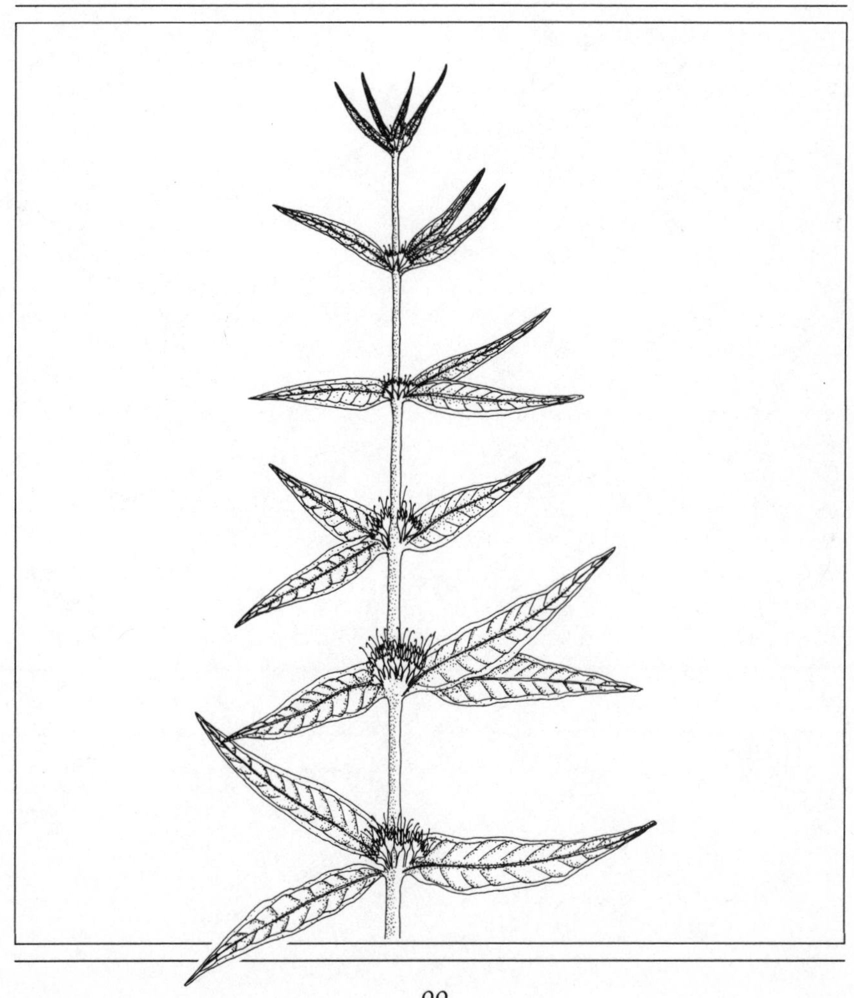

CREEPING resolutely over the watery eye of the bog, the swamp loosestrife forms a dense network of interlaced, arching stems. Bountifully endowed each summer with deep pink blossoms, this colorful shrub spreads rapidly into the shallows and hastens the succession from wet areas to dry land. Wherever a recurved, corky stem touches the water, air-filled, spongy tissue may develop. Swelling to four times its normal thickness, the stem becomes buoyant, develops roots, and forms a new arching stem.

The leaves of the emergent swamp loosestrife are lance-shaped and opposite. Throughout the summer, bell-shaped, magenta flowers will be developing in the leaf axils on the upper stems. By fall round fruit capsules will have formed on the plant. The arching branches have light brown bark that begins to flake as they get older.

Poison Sumac

Rhus vernix

DESCRIPTION

Classification: Shrub
Size: Height to 25′
Characteristics: Deciduous; 7–13 alternate, compound leaflets; hairless twigs; smooth, gray bark; greenish blossoms; white berries; flowers June–July; *entire plant poisonous to touch*

HABITAT

Swamps, bogs

RANGE

Maine to Florida; west to Texas; north to Minnesota

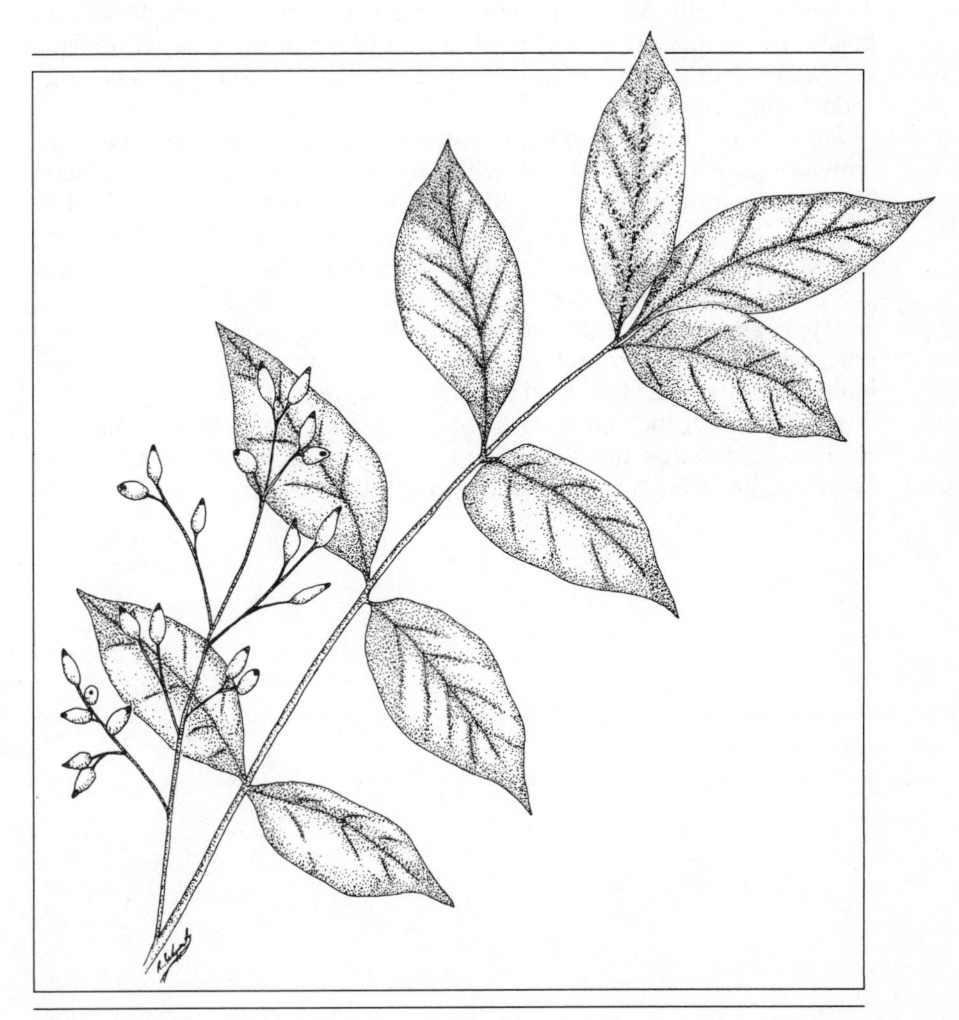

STANDING amidst the muted browns, pale yellows, and somber grays of the wetlands in autumn, the brilliant scarlet foliage of the poison sumac startles the eye. Concealing its blistering character, the colors lure the amateur naturalist to the plant for an inevitable encounter. As the leaves and white berries are collected, the oil exuded by the plant contacts the skin and produces an unpleasant itching and swelling.

The later formation of oozing, painful blisters, and the general discomfort caused by the plant on the unwary are not soon forgotten. "Leaves of three, let it be; berries white, take flight" now takes on an added dimension.

The derivation of the plant's scientific name is very appropriate. *Rhus* means "to flow" and *vernix* means "varnish." Poison sumac exudes a heavy sap containing a nonvolatile oil that is much more virulent than that of poison ivy, and the shrub is often considered one of the most dangerous North American plants. Treatment for exposure to either poison ivy or poison sumac involves scrubbing of the area with soap and water. Next alcohol is swabbed over the affected skin and a baking soda paste is applied.

Rhus vernix, also called poisonwood and swamp sumac, forms a bushy crown of gray, coarse branches which are reddish-tinged when young. The flowers form in loose, greenish clusters and will appear before the leaves attain their full size. White, waxy, one-seeded berries will be produced from the flowers. These fruits, forming on lone racemes that grow off the sides of the branch, are nontoxic to birds and other wildlife and will persist long into the winter. Providing food for bobwhites, pheasants, grouse, and rabbits, the sumac is an important source of nourishment when other food is generally scarce.

Economically, the genus *Rhus* yields a black varnish made from the clear, toxic sap that turns dark upon exposure to the air. *Rhus vernix* however, has a limited value in this respect.

Black Willow

Salix nigra

DESCRIPTION

Classification: Shrub
Size: Height to 120′
Characteristics: Deciduous; dark brown, deeply furrowed bark; alternate finely toothed leaves, green above, paler below; broad open crown; flowers April–June

HABITAT

Swamps, marshes

RANGE

Eastern half of the U.S.A.; scattered from west Texas to northern California

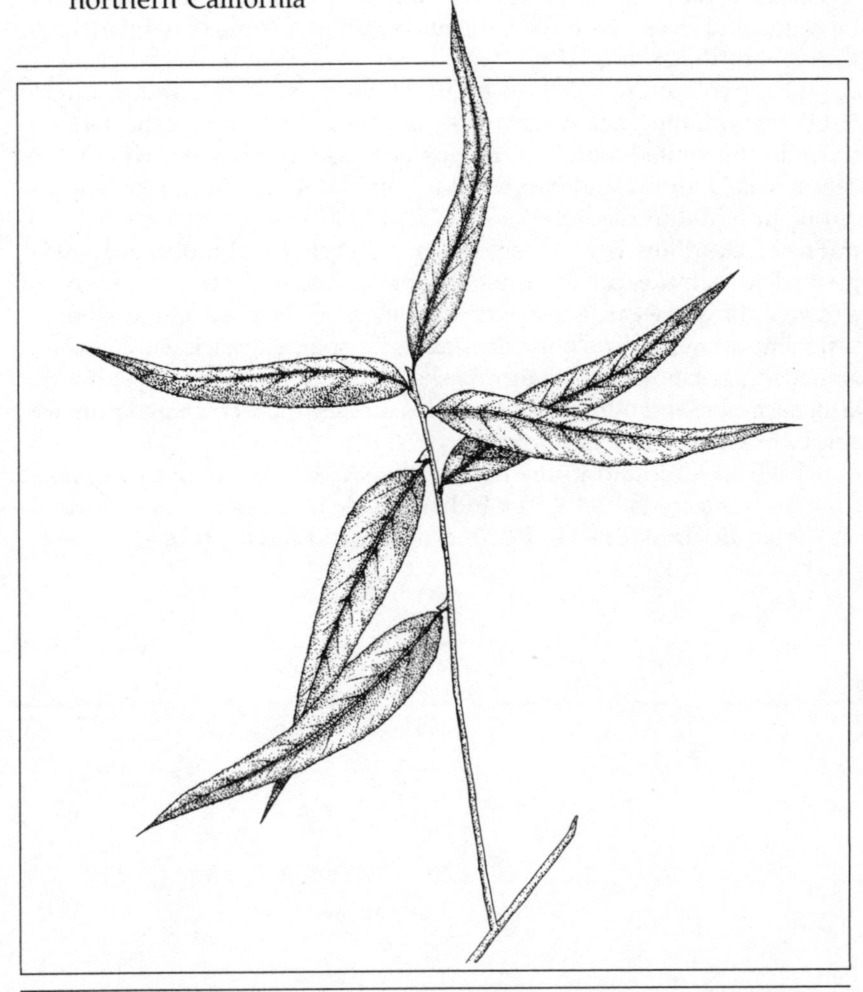

AS the warm sun increases the length of its daily visits, the cascading branches of the black willow begin to blush with a soft amber tinge. Soon the boughs are bursting with tiny yellow blossoms and resemble the magnificent golden mane of a lion or the long, lovely tresses of a lady. Alive with bees drinking the sweet nectar of the flowers, and with birds singing cheerfully, the willow's green buds swell with each passing day, finally exploding into a splendid leafy display.

The black willow prefers moist soil, where it often forms dense thickets. This shrub may attain a height of one hundred twenty feet under optimum conditions, but usually it stands from ten to forty feet tall. Its alternate leaves, lance-shaped and finely serrated, are light green and lustrous above with paler undersides. The twigs are reddish brown, slender, and smooth and offer sharp contrast to the dark brown, deeply furrowed bark of the trunk.

Opening from April to June, the yellow blossoms are pollinated mainly by wind and bees. Tiny fruit capsules are then formed, which ripen throughout the spring.

Of the two hundred to three hundred willow species found throughout the world, the black willow is the largest and has the most extensive range in the United States. In the lower Mississippi Valley, where it is commercially grown and harvested for timber, it often achieves heights of one hundred forty feet.

The black willow is very useful as a soil anchor and roots easily. It is planted to help keep stream bank erosion under control. The wood, although soft and weak, is also economically valuable. Used where strength is not important, it does not warp, crack, or splinter. Black willow wood is made into bats, boxes, crates, baskets, wicker furniture, toys, doors, artificial limbs, and pulpwood. In pioneer times, the wood was a primary source of charcoal for gunpowder.

Salicylic acid, found in the bark of the willow tree, derives its name from the genus name *Salix*. The Indians made tea from the willow wood and used the drink to soothe arthritic pain and reduce fevers.

Bald Cypress
Taxodium distichum

DESCRIPTION

Classification: Tree
Size: Height to 150′
Characteristics: Deciduous conifer; flat, thin, light green needles; gray to reddish brown bark; tapered trunk; pyramidal crown

HABITAT

Swamps

RANGE

Southeast and lower Mississippi Valley

STRETCHING skyward with an imperial air, the bald cypress is the unchallenged sovereign of the southern swamp. The elegant drapery of Spanish moss is smugly displayed as the cypress towers above its comrades, the oaks, willows, and tupelos. Patriarch of the southern wetland, *T. distichum* thrives through the centuries unscathed. Individuals, each hundreds of years old, still flaunting the virility of youth, dominate stands of bald cypress. A number of trees have been identified in recent years that were mere saplings when Columbus pondered the Genoa horizon.

Probably the most striking feature of a bald cypress stand is the peculiarity of buttress roots. Peeking above the water's surface, like periscopes from a fleet of submarines, these knobby extensions of the basal root system serve a dual function. Primarily, the buttress roots extend the tree's anchorage in the soft and capricious soil. Further, air exchange for the cypress is often limited as the wetlands swell with spring rains. The main roots, barred by rising waters from direct contact with air, are rendered ineffective in this service. The buttress roots, which may project several feet above the water's surface, assume the air exchange function.

The pale green leaves, or needles, of the bald cypress are usually less than one inch in length and are held laterally on short stems, or branchlets. This is a cone-bearing but deciduous tree, and when the autumn leaf drop occurs, the entire branchlet, similar in structure to a compound leaf, falls. Both the staminate and pistillate cones are small and roughly spherical. The male cones grow in loose, purple clusters of three to four, while the female cones are held singly. The winged seeds are quite tiny, often smaller than one quarter inch, and require well-saturated soil to germinate.

The bald cypress is a valuable tree in the wetlands habitat. While its contribution as a soil anchor is obvious, it often serves the biotic community in more discreet ways. From its deepest root to the tip of its crown, the bald cypress offers protection, food, and nesting sites to innumerable swamp organisms. The woodpecker's endless search for embedded insects creates cavities used by many bird species for their nests. The crown itself teems with flying insects and the birds and frogs who prey upon them. Within the torpid waters, minnows and the fingerlings of game fish dart through the cavernous disarray of entwining roots evading the snapping jaws and bills of predatory reptiles and birds.

The value of the bald cypress also extends to the economic practicalities of human life. The wood's fine grain results in a pleasing finish for cabinets and interior trim. As the bald cypress ages, the tree's core often disintegrates, leaving a hollow center. Such a tree is still capable of survival. A hollow tree is not a desirable lumber product and will be bypassed by loggers.

The bald cypress has weathered the cataclysmic disruption of past geologic eras and stands today as a venerable symbol of survival. A more rigorous test is yet to come as man expands his technological capacities.

Yellow Birch

Betula lutea

DESCRIPTION

Classification: Tree
Size: Height to 100′
Characteristics: Deciduous; lustrous, silvery gray bark; paired leaves to 4″

HABITAT

Swamps

RANGE

Maine to North Carolina; west to the Great Lakes

AS winter slowly eases its icy grip, a critical time for wetlands songbirds begins. Rusty voices strain, giving flight to melodies of yet another spring. The still frozen ground remains unyielding to worm-seeking beaks, and most insects linger in their winter coma. A sudden spring snowfall conceals any last vestige of berries and seeds. Many birds might face a bleak future except for the yellow birch. The female cones of *B. lutea* disintegrate so slowly through the winter that many seeds are just being released as spring approaches. It is not unusual to see these dark seeds dotting the surface of a fresh snow. Thus, the yellow birch provides a vital eleventh-hour food source for many wetlands birds.

Male and female flowers, borne on the same tree, blossom as the spring foliage unfurls. The male flowers are clustered on three-inch catkins dangling gracefully from short stalks. The female flowers develop in smaller upright cones, which will hold the seed through the piercing days of winter.

The bright lustrous bark of the yellow birch readily peels into crisp, brittle curls. Highly flammable even when wet, these bark curls make excellent kindling for campfires. The bark of new growth suggests the yellowish hues that named the tree, but with maturity the color evolves to silver gray and finally a rich chestnut.

Yellow birch lumber is the most desirable of all the birches. The close grain and strength of this wood creates an excellent product for flooring and cabinetry.

One of the largest eastern deciduous trees, the yellow birch is most commonly found in moist, rich soil. It may be one of the pioneer trees after a clear cut or burnout but is also a representative climax species in a mature wooded swamp.

Atlantic White Cedar

Chamaecyparis thyoides

DESCRIPTION

Classification: Tree
Size: Height to 80′
Characteristics: Evergreen; scaly leaves in fanlike sprays on flattened branches

HABITAT

Swamps, bogs

RANGE

Eastern and Gulf coasts

THE Atlantic white cedar, not to be confused with arbor vitae, or *Thuja occidentalis*, is part of nature's wetlands high-rise complex. The tall, slender trunk supports a conical crown where tree frogs, birds, and many small mammals conduct their daily business. In southern regions the dense growth of the cedar is interlaced with that of its companion tree, the bald cypress. There the forest canopy is draped with the entangled filaments of Spanish moss and the tree-loving vines of greenbriar.

White cedar flowers are borne deep within tiny, bluish cones. Brown winged seeds, developing after pollination, are shed from the cones in the autumn and are dispersed by both wind and water.

The reddish brown trunk conceals an inner bark that can be peeled in long fibrous strips. Primitive people discovered this to be a useful weaving material and fashioned baskets and other containers that could hold water. The roots of the white cedar thrive in acid swamp waters, especially those offering a sandy subsoil.

The wood of the white cedar is soft and fragrant. Saturated with natural fungicidal chemicals, the wood is rot- and insect-resistant. For this reason, it has proved to be an excellent building material enjoying both exterior and interior uses. Long popular as home siding, cedar shingles are increasingly valuable today. Many New Jersey bogs contain aged submerged cedar logs, which are being mined for their commercial merit. Today white cedar is extensively planted as an ornamental.

The colonists cut the heartwood from white cedar logs and used them for sewer pipes. One historical account lists white cedar as possibly the first native wood to be used in the carving of organ pipes.

This beautiful conifer is known as a pioneer tree. Where storm damage, fire, or a logging operation has cleared an area of tree growth, the white cedar is often the first to be reestablished. The tender saplings meet with much adversity as they attract rabbits, mice, and deer who browse on the tasty tissue. Those trees which do survive will dominate the swamp for several years as other species become established. The cedars are easily outgrown, however, and are commonly shaded out by the taller black gum.

Swamp Cottonwood

Populus heterophylla

DESCRIPTION

Classification: Tree
Size: Height to 100′
Characteristics: Deciduous; leaves alternate on long petioles, broadly ovate, dark green above, pale green below.

HABITAT

Swamps

RANGE

Connecticut to Georgia; west to Louisiana, Kentucky, and Missouri

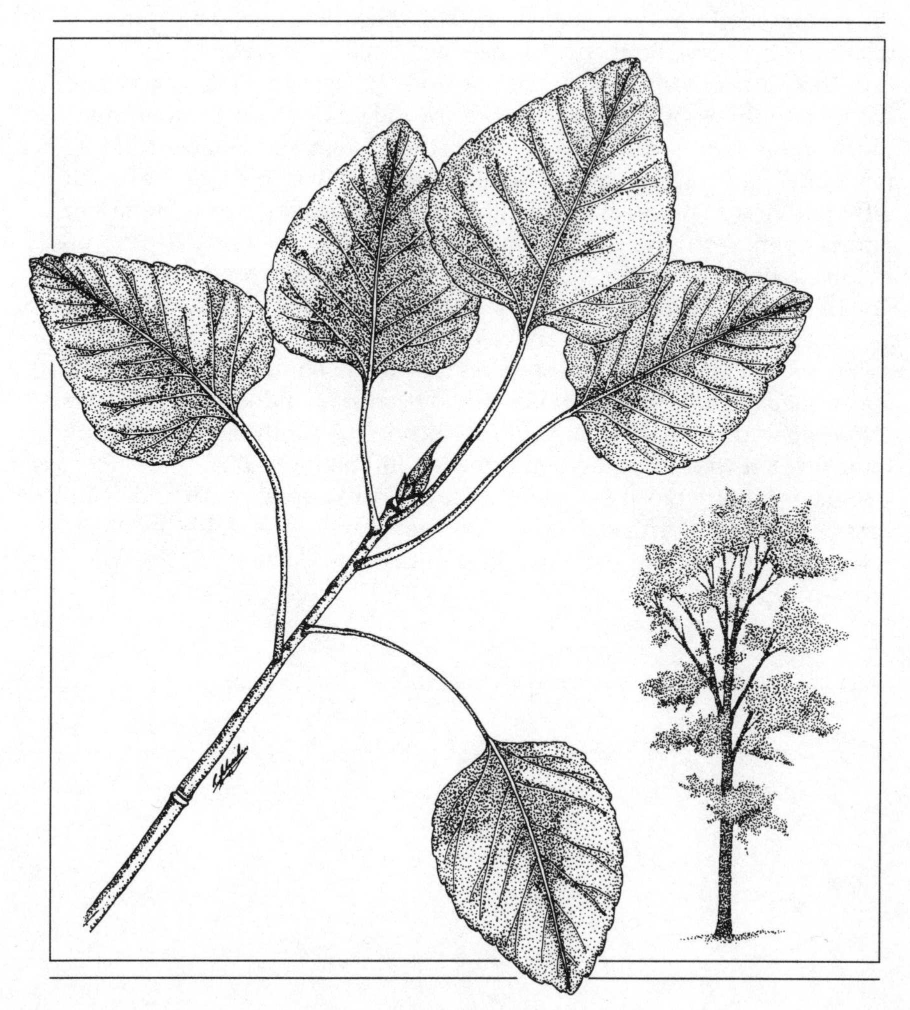

THE swamp cottonwood is a tree that offers hospitality to hundreds of wetlands species. Its broad, open branches offer a multitude of nooks and crevices appropriate for the nests and homes of varied wildlife architecture. For those seeking but a brief respite from endless chores or lengthy travel, the cottonwood extends an open invitation. As a temporary rest area or permanent residence, the dense greenery lends an ideal cover to all.

A type of poplar, the cottonwood begins to flower each spring before the leaf buds begin to open. The cottonwood is a dioecious tree, meaning that the male and female flowers are borne on different plants. Upon proper pollination, the fruit sets into tiny reddish brown capsules. Bearing the long, drooping strands of fruit, the female trees are easily distinguished from the males. As the tree begins to unfold its leaves, thousands of bristled seed capsules are released.

Soil that is well flooded for the better part of the year is ideal for the cottonwood's growth. Attaining heights of one hundred feet in the south, the cottonwood rarely reaches fifty feet in New England. The light reddish brown bark is marked with narrow, shallow fissures.

Although a closely related cottonwood, *P. deltoides*, is an important source of pulp wood, the swamp cottonwood enjoys little economic use. The wood is light, soft, and quite weak, therefore not a valuable building material. Deer, rabbits, and rodents, however, would suggest a much different view. The seedlings, twigs, bark, and leaves are all important nourishment for these wetland mammals.

Throughout American history, the swamp cottonwood has been a favored medicinal plant. The sticky leaf buds, mashed in lard, were used as a soothing salve for sunburns and minor skin inflammations. Similar salves were also rubbed on the chests of those suffering from coughs and upper respiratory infections. An expectorant and antiseptic gargle was prepared by soaking the buds in alcohol. A soothing tea or spring tonic was brewed from the buds steeped in boiling water.

Boiled roots were used as poultices to treat sprains, rheumatism, scurvy, gout, and arthritis. And as if that were not enough, even the water that collected in cottonwood crevices was claimed to have wart-removing properties.

Larch

Larix laricina

DESCRIPTION

Classification: Tree
Size: Height to 100′
Characteristics: Deciduous conifer; 1″ pale green needles borne in clusters

HABITAT

Swamps, bogs

RANGE

Maine to Pennsylvania; west to Illinois

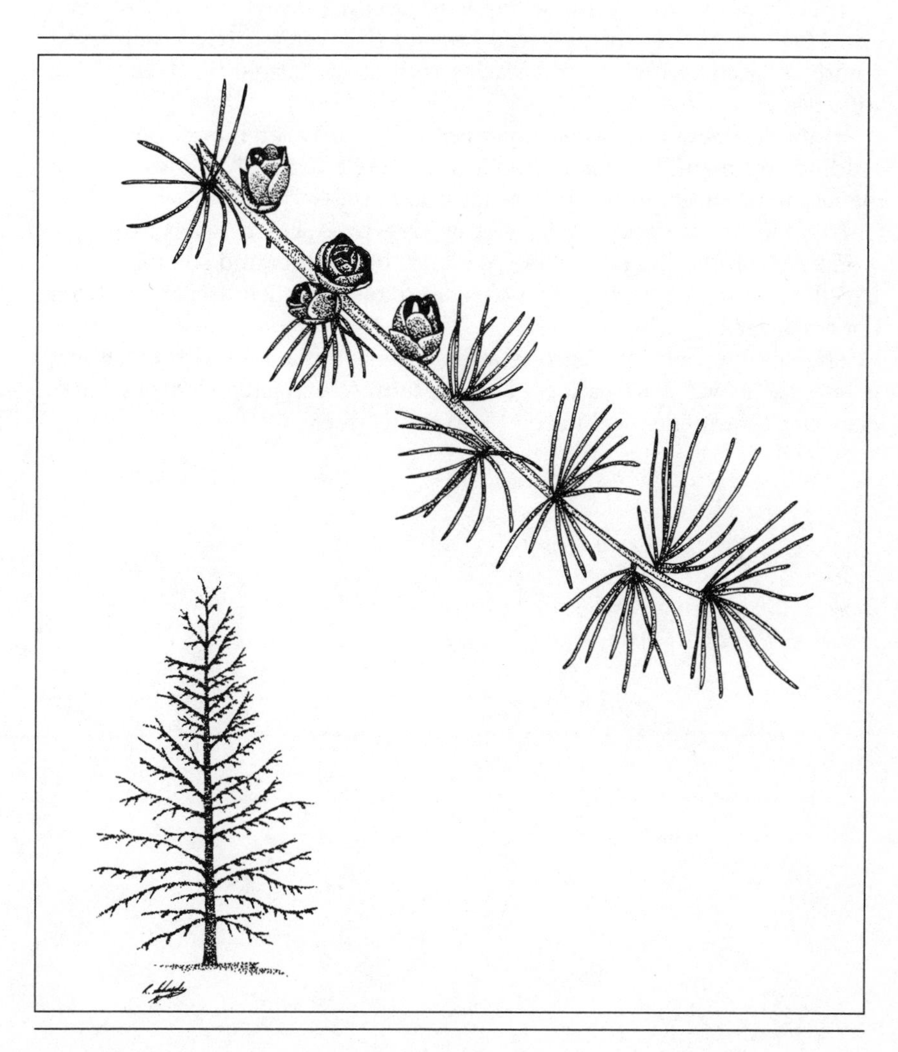

AUTUMN is a time of crisp, clear days and trees ablaze in their most splendid dress. It is a time that most conifers retire into the shadows to await their moment of glory, when a bleak winter landscape is gratefully broken by their deep evergreen tones.

For the larch, or tamarack, a deciduous conifer, autumn is indeed the choice season. The feathery illusion of gracefully leaved branches will quickly fade as brisk autumn breezes hasten the shedding process. As the chlorophyll fades in its soft, pale green needles, sunlight seems to sprinkle the tree in golden tones.

The tiny, chestnut brown cones, once bearing the flowering parts of the larch, spread their scales and free minute winged seeds. Grouse and other birds will feed upon the seeds throughout the winter. The cones will cling tenaciously to the branches as winter winds stress their grip.

The versatile larch is no stranger to stress, however. Larches grow shrubby and greatly stunted well beyond the Arctic Circle. Intermixed stands of larch and white and black spruce grow farther north than any forest tree.

In the temperate regions, however, the sturdy larch is of benefit to wildlife and mankind alike. Grouse and rabbits regularly browse on the leaves, while deer prefer the tender branches.

A common wetlands species, this tree produces a heavy, durable wood resistant to the degenerative effects of weather and soil exposure. Excellent fence posts, utility poles, and railroad ties are hewn from tamarack logs.

The shallow roots are strong and fibrous. Indians used these fibers to sew their birch bark canoes. Finally, tannin, extracted from the bark, is an important chemical used to process leather.

Silver Maple

Acer saccharinum

DESCRIPTION

Classification: Tree

Size: Height to 100′

Characteristics: Deciduous; leaves five-lobed, opposite, deep green above, silvery below

HABITAT

Swamps

RANGE

Eastern half of the U.S.A.

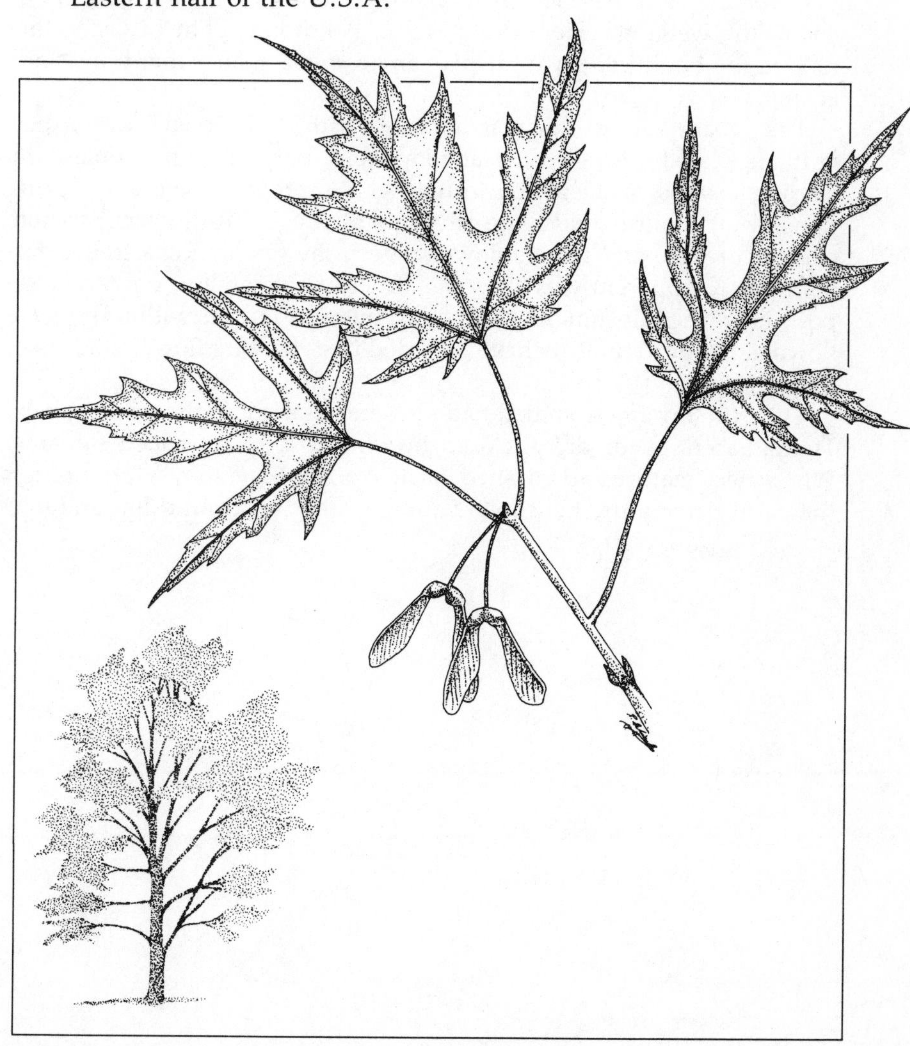

OCCASIONALLY man has encountered a fellow organism that seems to challenge his pursuit of exploiting the natural world. Such an organism is the silver maple. Attempts to collect the sap for maple syrup have been less than successful. While the tree does, indeed, produce the sweet fluids, their volume and richness are not worthy of the effort.

The wood of the silver maple has been harvested for lumber. Yet it is a "soft" maple of only medium hardness and far too brittle for fine crafts.

And finally the silver maple has been planted as an ornamental shade tree. But here, too, the tree enjoys only partial success. The brittle branches are readily pruned by brisk winds, and the massive volume of winged seeds further contributes to the litter.

Silver maple is best left in its natural environment, the rich, moist soil of the wetlands. Here the graceful, pendulous branches, nodding toward the bountiful earth, display an aesthetic quality that transcends economics.

The broad, round crown is amassed with dark green leaves, their bottoms plated with silver. Dangling from long crimson petioles, the sterling leaves sway elegantly with the gentlest breeze. Before the spring rains swell the leaf buds, the silver maple begins to flower. Separate clusters of male and female flowers adorn the tree in a chartreuse garnish. There apparently exists some structural variability, however. Reports have been documented of both stamen and pistils within the same flowers. Other naturalists have recorded trees bearing flower clusters of only one sex.

The fruit forming from pollination is characteristic of the maple genus. Two adjacent seeds are enclosed in winged capsules called samaras. When these mature and are shed in late April or May, they whirl through the air like miniature helicopters approaching the ground for landing.

Swamp White Oak

Quercus bicolor

DESCRIPTION

Classification: Tree
Size: Height to 100′
Characteristics: Deciduous; leaves alternate, pale olive green, smooth above, slightly fuzzy below

HABITAT

Swamps

RANGE

Maine to South Carolina, west to Minnesota and Arkansas

HIGH in the crown of a swamp white oak, a gray squirrel nimbly leaps from branch to branch. The September air alerts the busy animal to the approaching colder season. It is a time of preparation. The sweet acorns of the swamp white oak are ripening and are already dropping from this broad, sturdy tree. Soon other wetlands creatures will compete with the squirrel for this premium food. White-tailed deer, turkeys, and small rodents will scavenge the acorns littering the forest floor. Those who can take flight, mallards, wood ducks, and woodpeckers, will also diminish the squirrel's share of the fresh, tasty acorns that are held out of the reach of the ground dwellers.

Although not a common tree, the swamp white oak usually generates large local populations. While it occasionally will be found in pure stands, it is more likely to be encountered with sweet gum, silver maple, sycamore, and willow. *Quercus bicolor* is a hardy wetlands tree, often living more than three hundred years. It prefers the heavy, poorly drained soil of lowlands, swamps, and marsh edges. The color of the deeply furrowed bark tends toward an almost nondescript grayish brown. Along the broadly spreading branches, the bark is rather ragged and sheds in large flakes. The result suggests a somewhat scruffy appearance to an otherwise lovely tree.

The pale olive leaves of the swamp white oak are among the larger of the oak leaves. Averaging six inches in length, these alternate leaves are smooth above and slightly fuzzy below.

Most individuals of the swamp white oak are at least twenty-five years old before they begin to flower. This tree is monoecious, meaning that the male and female flowers, while borne on different structures, are present on a single tree. Each May from two to five tiny pistillate flowers cluster on short spikes and yellowish green male flowers hang in slender, four-inch catkins.

The flowers are pollinated and seeds develop through the summer, forming edible and even quite tasty reddish brown acorns. Approximately one inch long, the acorns are usually held in pairs on rather long stalks. By fall, most of the acorns have matured; they will drop from the tree by the first frost. Typical of most oak species, the swamp white oak produces an unusually heavy crop of acorns every three to five years.

In the form of planed lumber, the wood of the swamp white oak is fairly indistinguishable from white oak and is often marketed as such. It is strong, light brown, close-grained wood that is excellent for furniture, cabinets, and interior trim. It was once an important source for barrel staves, but that use has diminished along with the demand for wooden barrels.

Water Oak

Quercus nigra

DESCRIPTION

Classification: Tree
Size: Height to 80′
Characteristics: Deciduous; slightly lobed, spatulate leaves; dark gray bark; acorns single or paired

HABITAT

Swamps

RANGE

New Jersey to Florida; west to Texas and Missouri

THE stately water oak, sharing the southern swamplands with such species as willow, sweet gum, and swamp red oak, provides vital food and shelter for a variety of birds and mammals. The wetlands natives enjoy the tree's annual cycle of vegetative budding, blossoming, and seed production. The large, round acorns of the water oak are an important source of protein for such herbivores as squirrels, wild turkeys, and certain species of ducks.

Shiny, blue-green leaves hang from rather short, flattened petioles. The leaf shape is quite variable, although the three-lobed spatulate form is most common. These, as well as variously lobed or even complete leaves, may occur on a single tree. The foliage is smooth and shiny on both sides, while the hue is somewhat subdued on the underside. Retaining their color well into the fall, water oak leaves finally turn yellow before dropping in early winter. Overall, the tree's graceful form and lovely rounded crown have made it a popular shade tree for streets and parks throughout its range. For contrary to the implications of its name, the water oak does not require footing in permanently submerged soil.

Shortly after the tender spring leaves begin to catch an April breeze, the flowers blossom. Pollen-bearing blooms hang in two- to three-inch-long red catkins. The female flowers, held on short, downy stalks, will soon begin the development of the acorns.

Water oaks grow from large, nearly round acorns that are held on the parent tree for two years before they fully ripen and are shed. The nuts are yellow-brown during the first season but darken considerably by maturity.

The wood of the water oak is hard, strong, and close-grained. It is probably the most valuable hardwood in the south, although it is not a particularly high-quality wood. Generally marketed as red oak, the lumber is used for flooring, furniture, and railroad ties. In moist soil, a sapling will be ready for timber harvesting within seventy years.

Freshwater Clam

Sphaerium sp.

DESCRIPTION

Classification: Mollusk

Size: Diameter to ½″

Characteristics: Thin, shiny, yellow-brown shell

HABITAT

Swamps, marshes

RANGE

Continental U.S.A., east of the Rockies

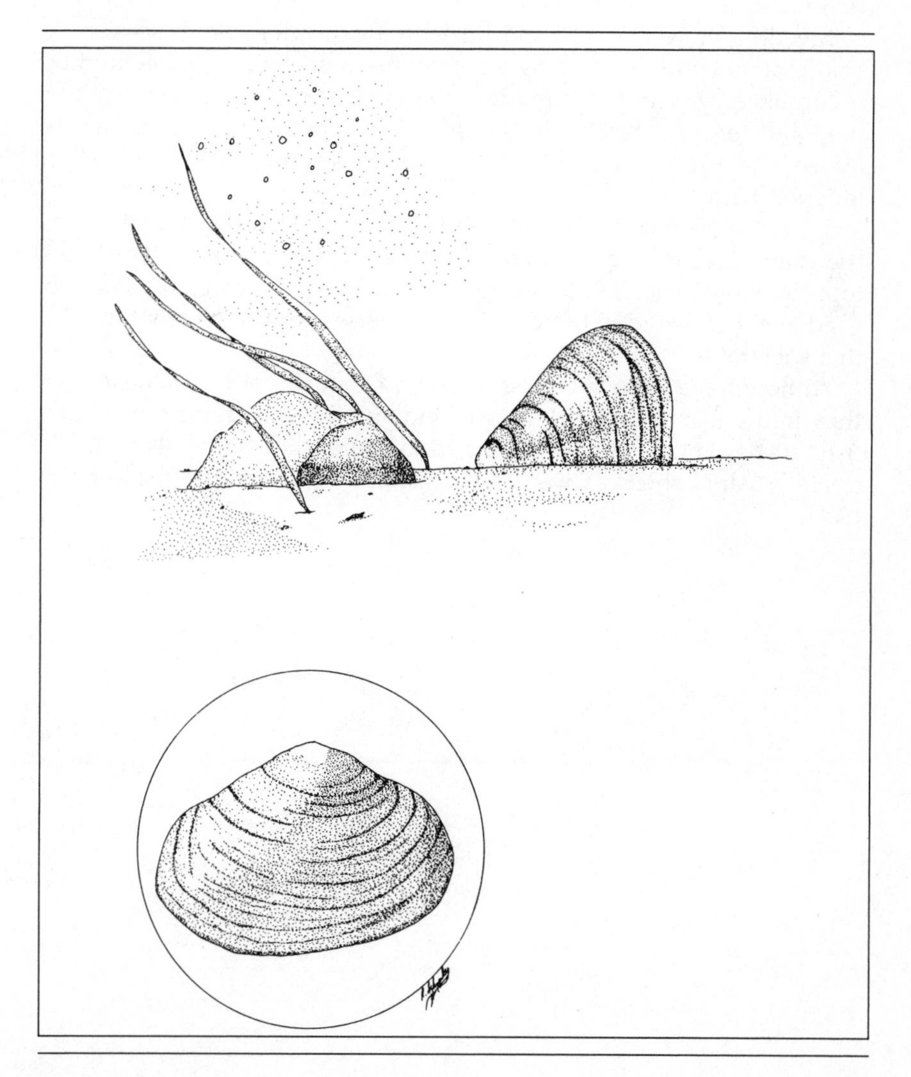

AS sunlight penetrates the rippling water of a marsh, the translucent shells of tiny freshwater clams glitter below the surface. This translucent quality seems to capture the sun's rays and creates an illusion of gems haphazardly sprinkled among the vegetation. The weak, thin shells are marked with conspicuous ridges or growth rings. The precise composition of these shells varies with the availability of lime in the surrounding waters.

Clinging to submerged vegetation or simply secure on the wetlands muddy bottom, *Sphaerium* sp., unless disturbed, maintains an almost constant flow of water funneling through its body tissue. Mollusks like the freshwater clam and mussel extract nutrients from the waters they inhabit by a process known as filter feeding. In the clam, fleshy, teethlike tendrils surround the mouth of the siphon, which admits the nutrient-rich water.

Breeding between April and July, the freshwater clam lacks the glochidial stage common in many other mollusks. Instead, a limited number of fertilized eggs are nurtured in a specialized chamber near the gills of the adult females. Before the summer's end, the juvenile clams are released and drop to the water's bottom to resume their development independently.

With the approach of winter and the lowering water temperatures, the clams prepare for a period of reduced activity. The wedge-shaped foot, previously used for attachment, now becomes a digging tool to be used along the muddy bottom. Clams overwinter completely buried in this substrate.

Although the freshwater clams offer no direct economic value to man, they figure significantly in the wetlands food web. Animals, including fish, turtles, mink, and raccoons, draw important minerals from the easily crushed shells as well as body-building protein from the fleshy tissue.

Freshwater Mussel

Lampsilis siliquoides

DESCRIPTION

Classification: Mollusk
Size: Length to 4″; width to 2″
Characteristics: Two oval, light or dark green, ridged shells hinged together

HABITAT

Swamps, marshes

RANGE

Continental U.S.A.

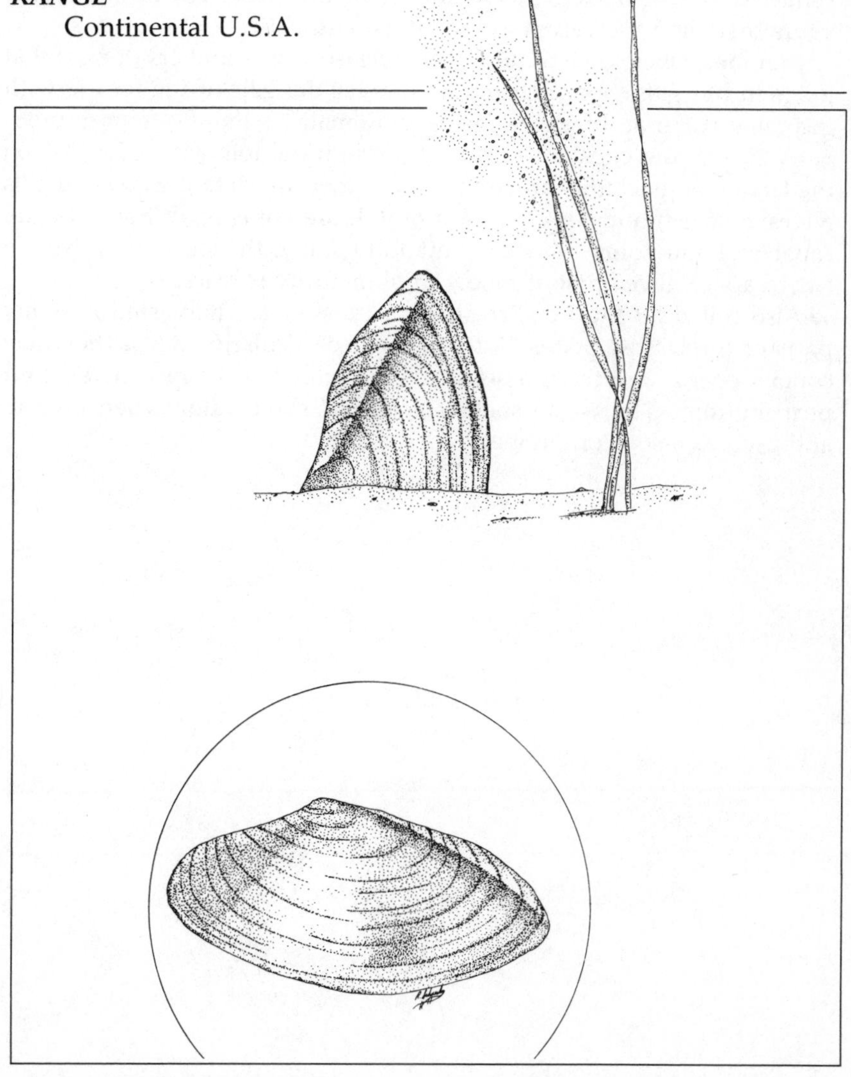

THE freshwater mussel, one of the wetlands' more sluggish inhabitants, is generally seen dawdling in the loose, muddy bottom of the still waters. Like that of most adult mollusks, its body structure curbs its mobility, so that its living state is often in question. Yet, in fact, the mussel is a site of nearly unceasing activity owing to its filter-feeding mechanism. Water is drawn in through the mouth of a siphonlike structure. Nutrients and dissolved oxygen are absorbed by the soft body tissue, and waste products are surrendered. The waste-laden water then exits through the second siphon opening in a continuous flow.

A single fleshy foot attaches the mussel to its substrate. This spadelike projection extends from within the hinged shell (bivalve) and can be retracted at will. Along the perimeter of the shells are tiny teethlike spurs that align the valves as they open and close.

In midsummer, each adult female release large numbers of eggs that lodge in her gills. Spermatozoa then enter the gills of the female with incoming water and fertilize the eggs. Remaining there safe from predatory forces, the eggs are nurtured through the following May. When the larvae, or glochidia, are finally freed, they attach to the fins and gills of resident fish and for close to a month are parasitic. When they are capable of autonomy, they drop off and fall into the loose mud. Nearly two years of development follow until maturity is reached.

Also called fat muckets, freshwater mussels are harvested in many parts of the United States. Although not particularly edible, they may contain pearls, and the mussel shells are valued as the raw material for pearl buttons. The waste shells are also of some value when ground and used as a soil conditioner.

Left-Handed Pond Snail

Physella heterostropha

DESCRIPTION

Classification: Mollusk
Size: Length to 1″
Characteristics: Color varies from pale yellow to black; whorled, pointed shell with opening to the left

HABITAT

Swamps, marshes

RANGE

Eastern half of the U.S.A.; related species throughout the U.S.A.

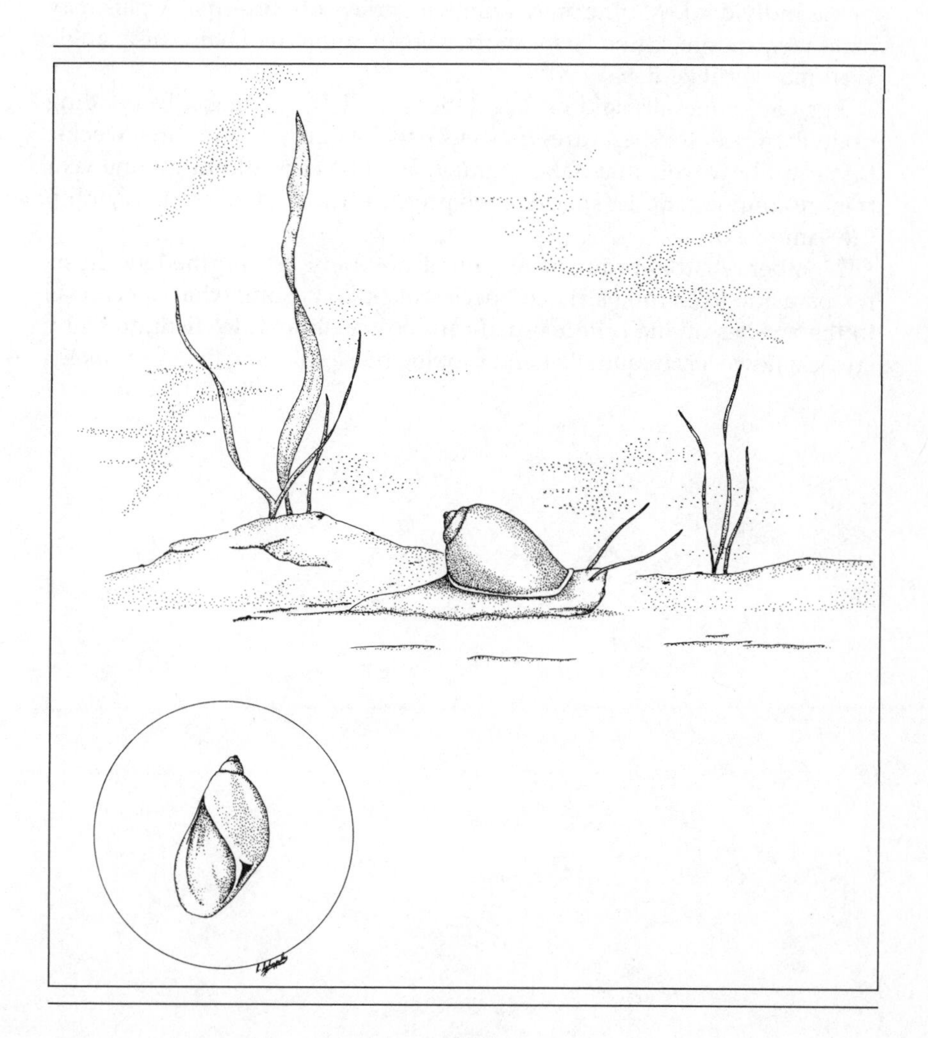

TO observe a pond snail gliding smoothly along vegetation, a beginning naturalist might never suspect the complexity of this familiar creature. A controlled series of muscular contractions offers pond snails such agility that they can even creep upside down along surface waters. They breathe through a lunglike cavity or exchange gases directly through their skin. Pond snails have been known to survive prolonged periods embedded in ice or enduring a drought.

Two nonretractile tentacles extend from the animal's head. At the base of each tentacle is a light-sensitive eye. The tiny tongue, like a carpenter's rasp, is covered with miniature, grating teeth. Feeding upon both plant and animal matter, pond snails have become standard aquarium scavengers.

An aquarium with a single pond snail, however, may soon be overrun. Like other members of the order Pulmonata, pond snails are hermaphroditic, meaning that both male and female organs appear on a single individual. Mating may follow a variety of patterns. A pair may mate reciprocally, three or more may create a mating chain, or a single snail may fertilize itself.

Eggs are generally laid on vegetation in a jellylike mass. Depending upon the water temperature, the eggs usually hatch within three weeks. Maturity, however, may take as much as two years, although one year is more common. Some species even produce a second generation within the same year.

Members of this order have gained notoriety as intermediate hosts for parasitic liver flukes. Each species of fluke is somewhat specific as to the species of snail. Pond snails are rather difficult to distinguish by species, however, frequently handicapping those studying the liver flukes.

Crayfish

Cambarus diogenes

DESCRIPTION

Classification: Crustacean
Size: Length to 5″
Characteristics: Pale carapace covering head and thorax; five pairs jointed legs, the first pair with pincers; two pairs antennae; jointed tail

HABITAT

Swamps, marshes

RANGE

Continental U.S.A., east of the Rockies

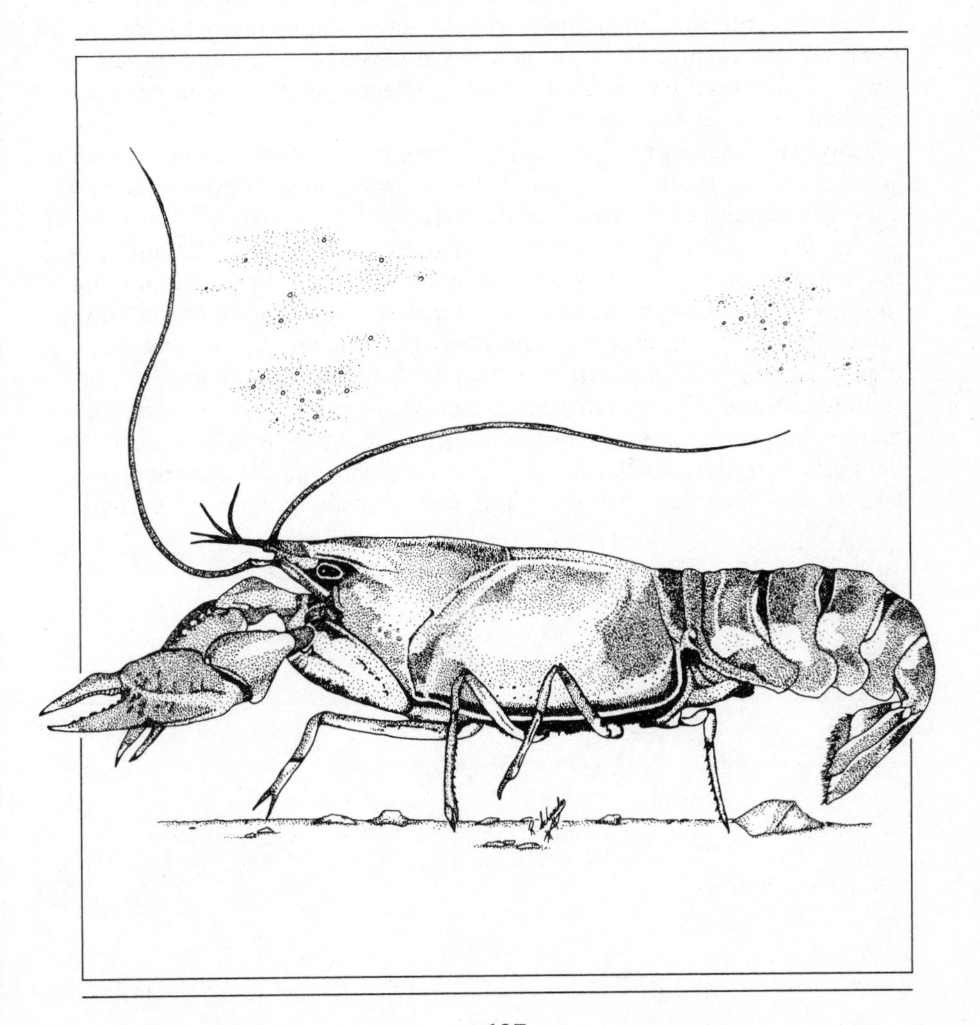

DESPITE its diminutive size, the common crayfish would make an intriguing model for an intimidating science fiction monster. Under the water's natural magnification, the pale translucent exoskeleton, the jointed plates, and the protruding eyes must create a threatening figure to smaller wetlands creatures. Four pairs of stiffly linked legs are used for walking, while the fifth and foremost pair is modified with menacing pincers normally used to hold and tear food. Though limited in size, these mini-claws are controlled by disproportionate muscle tissue and can inflict a painful pinch to a finger or toe. Two pairs of antennae provide sensory stimulation as they sweep the forward waters. The jointed tail ends in thin plates that can be fanned at will.

As though appearances were not enough, crayfish behavior contributes to the fiendish image. While the crayfish grows and strains within its carapace, the exoskeleton is periodically shed in total as the animal wiggles out the back. Ghosts of former lives litter the muddy bottom.

When startled or threatened, the crayfish escapes in backwards retreat. By contracting strong abdominal muscles and fanning its tail, a crayfish can shoot backward at amazing speeds. Limbs lost in battle or accident can be readily regenerated.

Crayfish usually prefer to eat vegetation but often are observed consuming animal tissue. They serve the environment as important scavengers, although they have been known to destroy nests of fish eggs.

Crayfish are very private animals. They usually hide by digging into the soft, spongy earth. They will venture out of the water, although not too far from the margin, and often tunnel along the water's edge building mud "chimneys" for air exchange. There they may remain concealed all day, emerging with the dark to satisfy their life-sustaining drives.

Indeed these animals have much to fear. To many fish, reptiles, and mammalian carnivores, crayfish are desirable prey. Man is an especially dangerous hunter. The Louisiana "crawdad" is a sought-after delicacy, and dead or alive, crayfish have long been popular laboratory animals.

Palamedes Swallowtail Butterfly

Pterourus palamedes

DESCRIPTION

Classification: Insect
Size: Wingspan 4–6″
Characteristics: Upper wings dark with yellow spots along perimeter and yellow crossbands; blue spot on each wing

HABITAT

Swamps

RANGE

Southern New Jersey through the southeastern states

GENTLY alighting on the spike of a pickerelweed, a palamedes swallowtail extends its coiled mouth tube into the depths of the graceful blue flower. As the butterfly greedily draws out the sweet nectar, sunlight glitters off its erect wings. Nutrient-rich, the nectar sustains the insect for yet another day.

The palamedes swallowtail is without a doubt the most distinctive butterfly of the southern wetlands. It inhabits every swampland of consequence from Virginia to Florida.

The moths and butterflies derive their family name, Lepidoptera, from the Greek words for "scaly wings." Indeed, these insects usually bear wings of scales layered in rows like shingles. The powdery matter that adheres to fingers when a butterfly is held is actually hundreds of tiny scales jarred loose from their slender attachments. The scales of the palamedes, like so many butterflies, are of two types. The first contains pigments that paint the wings in brown, orange, and yellow. Other scales contain no pigmentation but diffract sunlight to create the iridescent blues and violets.

Characteristic of its family, the palamedes swallowtail undergoes four life stages. Adult females lay their eggs, the first stage, on the stems or leaves of healthy plants. A wormlike grub, or larva, is hatched from the egg and eats the vegetation voraciously throughout this second stage. The larval form of the Lepidoptera is known universally as a caterpillar. The palamedes caterpillar is humpbacked and pale green with a pair of orange and black "eyespots" on its last thoracic segment. Caterpillars experience rapid body growth and generally enjoy four or five skin molts to accommodate their increasing size. Unlike most butterflies, the palamedes swallowtail may overwinter as a caterpillar.

The third stage is the pupa or chrysalis. Exuding a silken thread from its mouth, the palamedes caterpillar wraps itself as it attaches to a plant limb. The mottled green pupa is a resting period in the life cycle, during which the organism uses up much of the nutrition it stored as a larva. Most of the Lepidoptera survive the winter in the pupal stage. The fourth and final life stage is attained when the adult butterfly struggles free from the pupal case, spreads and dries its colorful wings, and flutters off in search of nectar, a mate, and the pleasure of free flight.

Butterflies are generally active during the daylight hours, and the palamedes swallowtail is no exception. They are attracted to brightly colored flowers such as the pickerelweed. Like many flowers that depend upon insects for pollination, the pickerelweed reflects the rich colors of the ultraviolet wavelengths. Most butterfly eyes, quite different in structure from human eyes, are sensitive to the ultraviolet hues.

Unless disturbed the palamedes swallowtail does not fly at night. Instead, swarms of adults roost communally on the highest branches of mature oak trees.

Green Darner Dragonfly

Anax junius

DESCRIPTION

Classification: Insect
Size: Wingspan 4″
Characteristics: Two pairs of wings, held horizontally when at rest; bright green head and thorax

HABITAT

Swamps, marshes, bogs

RANGE

Continental U.S.A.

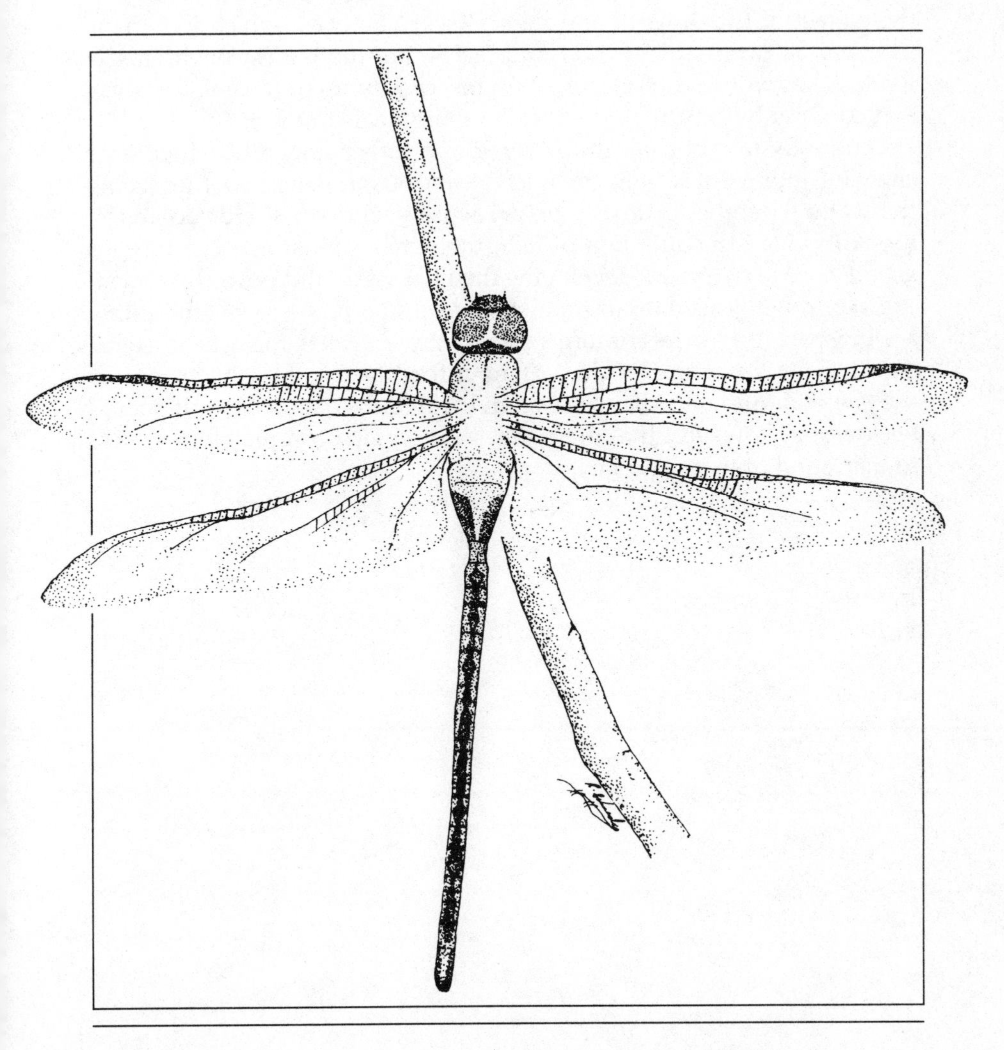

POISED on an emergent reed with wings outstretched, the green darner dragonfly rests briefly, then takes to the air once again. A familiar sight in a wetlands habitat, this darting splash of iridescence has maintained the secret of its flight for eons. Its thick, heavy body is too awkwardly structured to profit from Bernoulli's Principle, the physical phenomenon that explains the flight of birds and airplanes. Instead, a flight pattern that, for lack of a better term, has been labeled "unsteady aerodynamics" allows the dragonfly to streak through the air forward, backward, and sideways at speeds up to sixty miles per hour. The ease of rising into the air is measured by lift coefficient, the ratio of lift to wing surface. A single engine propeller plane generally has a lift coefficient of one, while a Navy jet fighter plane may have a lift coefficient as high as two. Dragonflies, using their "unsteady aerodynamics," can achieve a lift coefficient of six.

Sporting two pairs of rigid, transparent wings, the green darner is the largest of the dragon- and damselflies. The eyes, which may have as many as twenty-eight thousand facets, are the largest of the insect world. Dragonflies can observe their insect prey up to forty yards away and commonly capture and consume their meals in flight.

Dragonfly nymphs are dull-colored and rather ungainly. Their large chewing mouthparts help to satisfy voracious appetites as they comb the water's depths for insect larvae, worms, and small crustaceans. If they can avoid the attention of fish, their chief predators, the nymphs spend close to two years developing their body size and powerful wings.

Dragonfly fossils date to prehistoric times with one form suggesting a wingspan of two feet. More recently these insects have been called the "Devil's Darning Needle." Many a frustrated parent has repeated the tale of darning needles sewing up the mouths of bad boys. It should be stressed, however, that dragonflies are, in fact, completely harmless to man (and even to bad boys).

Mosquito
Culex pipiens

DESCRIPTION

Classification: Insect
Size: Length to ⅙″
Characteristics: Dark, slender abdomen; 1 pair narrow wings; 3 pairs jointed legs; long, thin proboscis

HABITAT

Swamps, marshes, bogs

RANGE

Continental U.S.A.

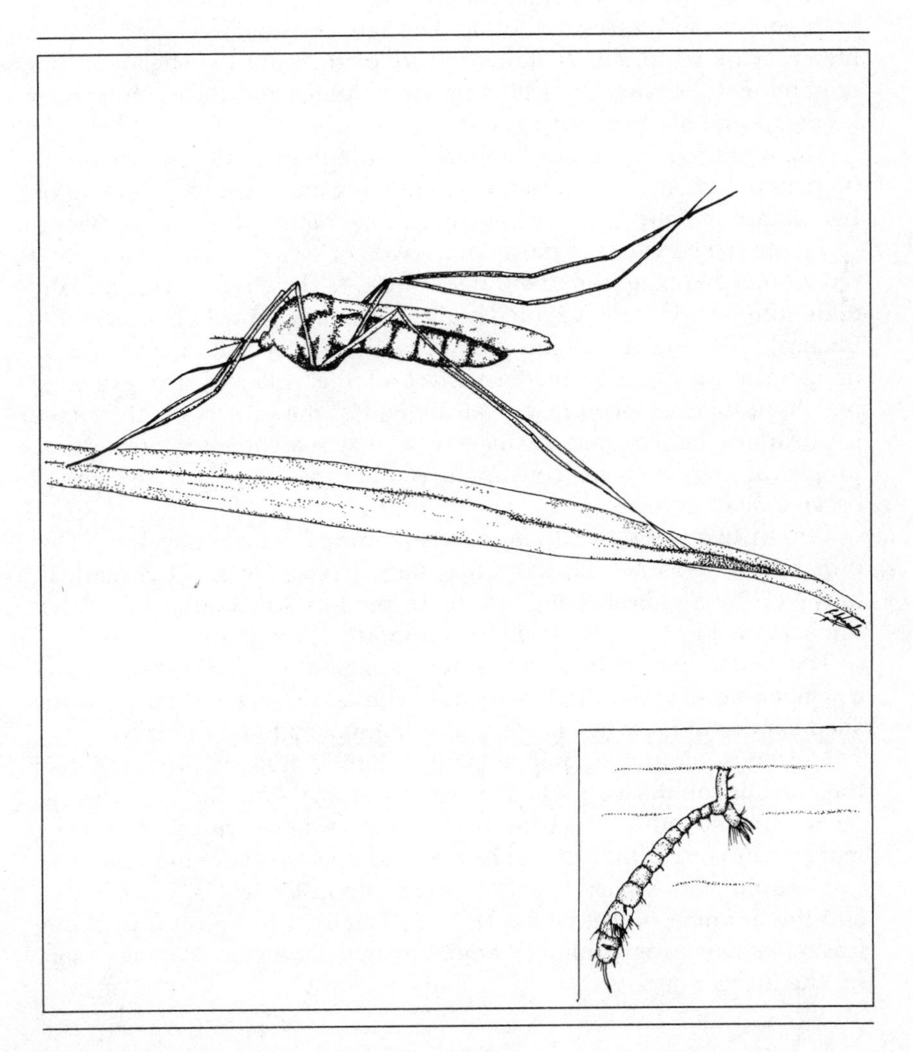

NO wetlands account could possibly be complete without mention of the famed vampire of the marsh. The soft drone filling the still air in early evening has unjustly labeled the wetlands as a hostile environment. Exposed faces and limbs are easy targets for the piercing mouthparts, as one's lifeblood is drawn out in short sips.

Indeed, the proboscis of the mosquito is unique among the order of flies to which the insect belongs. The sharp, spiked mouthpart of the female can penetrate the tough skin of most birds and mammals. Her appetite for animal blood seems almost insatiable.

Marked as "man's worst insect enemy," the various mosquito species are intermediate hosts to such diseases as yellow fever and malaria. *Culex pipiens*, commonly regarded as a pest, is also guilty of transmitting a parasitic round worm.

The diverse developmental stages of the common mosquito include both aquatic and terrestrial forms. Drought-resistant eggs and a short life cycle are important evolutionary adaptations for the shallow-water requirements of these insects. Temporary pools and fluctuating water levels are not life-threatening to mosquitoes.

The adult female's diet of animal blood enhances the proper development of her eggs. She lays them in floating masses, called rafts, upon the surface of still water. The aquatic larvae hatch within five days.

To the naked eye, the mosquito larvae, or wrigglers, appear as tiny red worms dangling from the water's surface. The wrigglers are actually quite complex. Their segmented bodies are equipped with hairlike tufts, tracheal gills, and the all-important breathing tube. The larvae rest by hanging at an angle from the surface of the water by the extended breathing tube. A lobed rosette at the end of the tube holds the larvae in position. The wrigglers are usually actively searching for food as they propel through the water consuming protozoans, algae, and suspended organic matter.

Within two weeks a curled, almost shrimplike form develops. This pupal stage does not eat; thus this stage is short-lived. It is marked, however, by a radical change in the respiratory function of the insect. The aquatic pupa displays two breathing tubes on its thorax.

The skin along the back of the pupa splits in two or three days, and a winged adult mosquito crawls out. The adult balances itself on the shed skin until its wings harden, and the mosquito takes flight.

Male mosquitoes are equipped with a slightly different proboscis than their female counterparts. Unable to pierce animal skin, the males subsist on nectar and other plant fluids. Both sexes have fringed rear wing margins and rest with their rear legs curved upward above the abdomen.

Mosquito control has largely centered upon larvicides, insecticides, and the draining of wet areas. The latter method has proved to be the least effective, as only minimal amounts of water are actually necessary for the insect's survival.

Fish, frogs, and countless invertebrates eagerly consume the plump wrigglers. Dragonflies and many songbirds depend upon adult mosquitoes as a major factor in their diets. Thus, the elimination of mosquito populations should be motivated by much more than man's inconvenience.

Brown Bullhead

Ictalurus nebulosus

DESCRIPTION

Classification: Fish
Size: Length to 20″
Characteristics: Black above, pale undersides; barbels; large mouth; no scales

HABITAT

Swamps, marshes

RANGE

Eastern half of the U.S.A.; introduced into California

FISH populations in the wetlands environment are limited by several physical parameters. The low level of dissolved oxygen resulting from rising water temperatures and reduced wind action does not provide suitable conditions for most game fish. Likewise, the shallower waters of swamps and marshes leave larger fish vulnerable to predatory herons and storks. As a result of these factors, most game fish remain in rivers, streams, ponds, and lakes. A notable exception is the brown bullhead.

A type of catfish, the brown bullhead is most at home along the muddy bottom of still, turbid waters. Tolerant of low oxygen levels, this hardy fish has reportedly survived extended periods completely out of water. It can also withstand much broader temperature variations than such finicky species as trout.

Catfish mate in late spring. Males carry the responsibility of nest building and guarding the developing eggs. Within a week, schools of coal black fry, protected by the male, eagerly explore their surroundings. As their whiskers, more correctly called barbels, lengthen and mature, the young fish become sensitive to the smells and textures of their aquatic world. The barbels are quite conspicuous sensory organs and contribute to the catlike appearance of the bullhead.

As the fish grows to maturity it may remain in the marsh, or if the waterways permit, it may swim out to a connecting stream, river, or lake. Bullheads are a popular game fish and a desirable panfish. Although averaging three to five pounds, they have been reported to reach well over one hundred pounds in deeper water.

Bullheads have very large mouths surrounded by the barbels. The smooth skin lacks scales and is dark to silvery in color. When handled, sharp spines on the dorsal and pectoral fins should be carefully avoided, as they can inflict painful injury.

Mosquito Fish

Gambusia affinis

DESCRIPTION

Classification: Fish
Size: Length to 3″
Characteristics: Silvery with dark-edged scales; vertical rows of spots on caudal fin; single spineless dorsal fin; dark eye bands

HABITAT

Marshes

RANGE

New Jersey; south to Florida; west to Texas

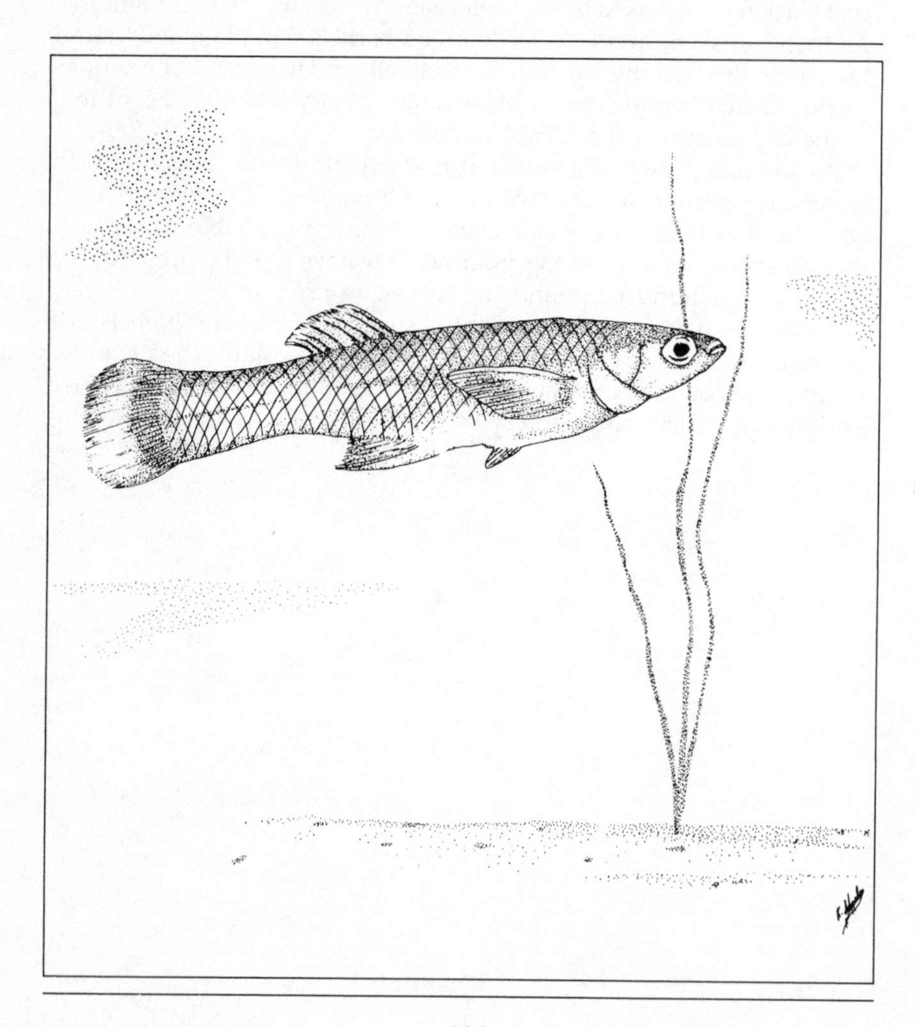

THERE is a saying among Caribbean fishermen that a man has fished for "gambusinos" when he has no catch to show for his day's efforts. Thus the diminutive mosquito fish derives its genus name from the Islanders' word for nothing.

From such humble origins, however, comes a three-inch minnow now recognized as a significant contributor to our wetlands environment. A single individual will eat its own weight in gnat and mosquito larvae daily. Native to southeastern United States, *Gambusia* have been widely introduced as an instrument of mosquito control. The voracious appetite, however, is often aimed at the broods of resident fish, and on occasion native populations have been eliminated.

Mosquito fish are prolific breeders capable of bearing up to two hundred live young at three-week intervals throughout the summer months. In the male the anal fin has evolved into an efficient copulatory organ. A single breeding often provides adequate spermatozoa for several broods. If the young are born in clear but highly vegetated water, their mortality is remarkably low.

Gambusia of all ages naturally fall prey to larger fish, particularly sunfish and young bass. Their tolerance to both fresh and brackish water and to a wide temperature range has permitted their introduction into many areas. Consequently, the mosquito fish has now become an important food source for various other fish species as well.

Although hobbyists make regular attempts to keep mosquito fish in aquaria, the fish is not considered a popular species. Aside from its rather indistinct coloration, the mosquito fish is quite pugnacious and is an aggressive fin nipper.

Bullfrog

Rana catesbeiana

DESCRIPTION

Classification: Amphibian
Size: Length to 8″
Characteristics: Largest frog in North America; green to olive back; cream to white belly

HABITAT

Swamps, marshes

RANGE

Continental U.S.A.

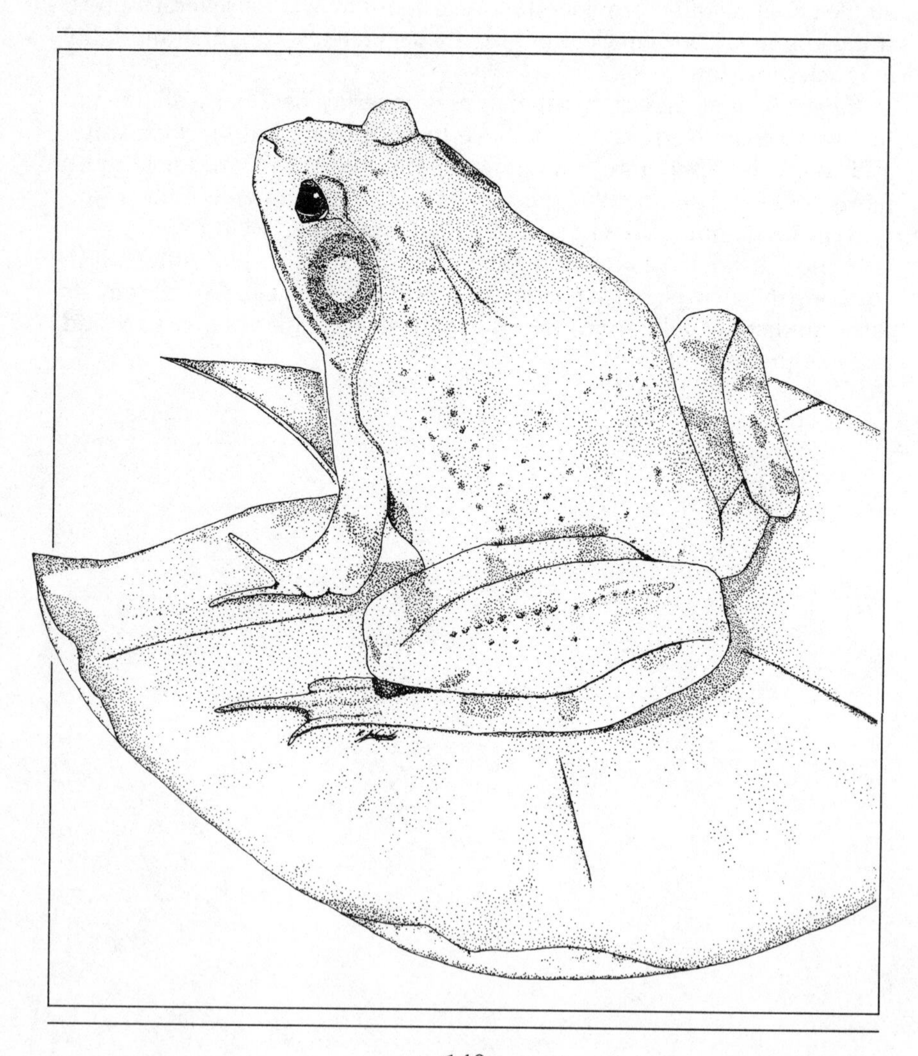

DEEP beneath the brittle ice, sheltered from subzero temperatures and a relentless wind, slowly beating hearts defy the winter tempest. Entrenched in the mud and ooze of the swamp floor, bullfrogs pass the winter in a near-comatose state of reduced metabolic activity. As plunging air temperatures nudge the frost line deeper, the frogs maintain their abbreviated existence unaware of the approaching peril. Should the frost extend into their hibernation site, death is a certainty. Fortunately, most swamp waters do not freeze to the mud line, and the frogs survive, emerging from their slumber as winter draws to a close.

Bullfrogs can be found throughout the country, preferring calm water bordered by a variety of emergent plants. There they prey upon small aquatic animals including insects, fish, and other frogs.

The bullfrog has a smooth, green head, an olive back, and dark bands on the hind legs. Over each eye and tympanum is a bony ridge. During the summer months, the males display a bright throat region.

Mating begins as early as February in the warmer states and extends through July as one moves northward. During this time, the males call with their very characteristic reverberating drum sound and await the arrival of a female. The mating process results in approximately twenty thousand fertilized eggs loosely attached by a jelly film.

Although the hatching of the eggs begins within days, another two years of growth are needed to fully develop the five- to six-inch tadpoles. Bullfrog tadpoles may be identified by their saffron undersides, spotted tails, and unusual size. Like most tadpoles their diet consists mainly of aquatic plants, although they have been observed consuming tiny animals as well. Within the following season the tadpoles complete their transformation to the adult form, but as many as five years may be required for full growth.

Unfortunately, the bullfrog is fairly easy to maintain in an aquarium and thus has become an exploited biological laboratory animal. This may be considered preferable, however, to its widespread use as a preserved specimen for classroom dissection activities.

The bullfrog is also subject to collection as a gourmet specialty. Frogs' legs offer a limited economic return, however, owing to the time needed to achieve marketable size.

Green Frog

Rana clamitans melanota

DESCRIPTION

Classification: Amphibian
Size: Length to 4″
Characteristics: Distinct ridges from the eyes approximately two-thirds of the way down the back; green to brown

HABITAT

Swamps, marshes

RANGE

Eastern half of the U.S.A.

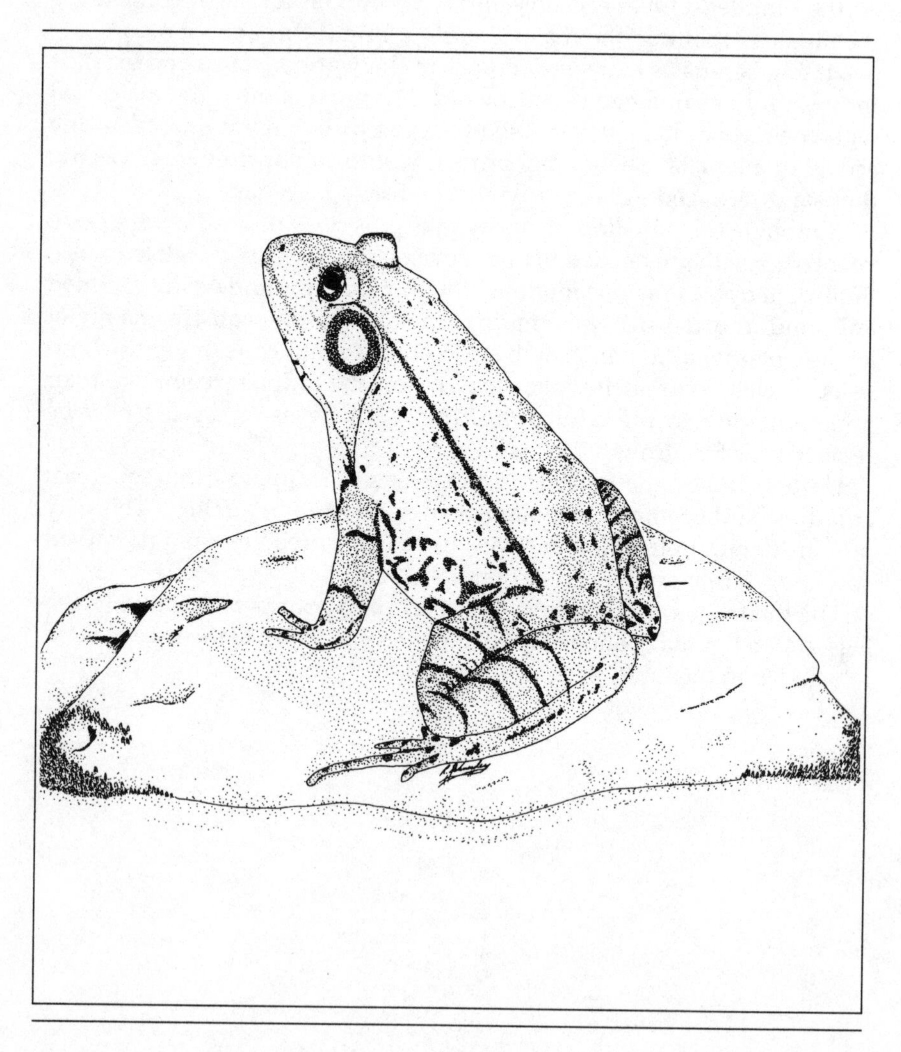

THE sedentary life of the adult green frog belies the frisky, frolicking antics of its youth. Darting eyes in constant search of flying insects and small aquatic animals and an occasional flicking of the tongue accounts for a major part of the day's activity.

The individuals of two subspecies constitute the abundant number of green frogs found throughout the eastern half of the United States. This small, gentle creature commonly inhabits quiet waters and, less frequently, nearby moist debris.

The upper body coloration ranges from green to brown. The underside of the frog varies from completely white to a pattern of dark flecks. The hind toes are moderately webbed.

In identifying sexes, the male's tympanum is usually larger in diameter than the frog's eye and is marked with a yellow center. During the breeding season the distinction is more obvious. It is then that the male's throat becomes bright yellow, while the female's throat remains white.

In early summer, the male initiates courtship with his low-pitched call. To produce this call, air is passed repeatedly from the lungs over the vocal cords to the vocal sac and back again. Since the mouth is kept closed, the resonating sound can be produced even while the frog is submerged.

Female frogs can recognize the call of their own species and easily find a mate. The couple remain united while the female lays her eggs, and the male releases his sperm over them. As many as five thousand fertilized eggs begin development near the water's surface, releasing tadpoles within five days. By the following summer, the tadpoles may reach a length of three inches and assume their characteristic green-tailed, brown-spotted, tawny-colored bodies. Enjoying a daily fare of oozes, algae, and other aquatic scums, the scavengers spend busy days wriggling about. It will be an additional three years before they complete their transformation into mature frogs.

Like the bullfrog, green frogs make excellent laboratory animals, both alive and preserved. The economic value of the green frog has been exaggerated by enthusiasts of specialty foods. While the hind legs are edible, their small size hardly justifies the sacrificed life.

Leopard Frog

Rana sp.

DESCRIPTION

Classification: Amphibian

Size: Length to 4″

Characteristics: Green to tan; rows of dark irregular blotches on the back

HABITAT

Swamps, marshes

RANGE

Continental U.S.A. except along the Pacific coast

AN initial encounter with a leopard frog will attest to the suitability of its name. Like a jungle cat camouflaged within the mottled foliage, it often startles the wetlands visitor with its sudden and often unexpected appearance.

There are four closely related species of the North American leopard frogs: *R. pipiens, R. spheriocephala, R. blairi,* and *R. berlandieri.* They were once thought to be a single species, but more recent studies suggest otherwise.

Leopard frogs range in color from tan to green. The color even varies within a single individual depending upon external conditions and the frog's state of mind. Upon capture the frog often loses its bright coloration and assumes a dull, nondescript hue. Adrenal and pituitary hormones control the expansion and contraction of the pigments contained in the skin cells, producing the altered surface appearance. The hind legs and underside normally remain their usual white color.

A distinguishing feature of leopard frogs is a pair of well-developed ridges extending the length of the back. A less obvious but equally significant means of identification is the pale moustache common in most species. Males usually exhibit an enlarged "thumb" and a fold of skin over the upper forelimb.

In the spring, leopard frogs seek marshy areas to breed. Female frogs prefer to attach their egg masses to submerged vegetation. Within three weeks, the eggs will hatch, and by the summer's end the pale yellow tadpoles will be approximately three inches long. The horny mouthparts are well suited to eating algae and the tadpoles will continue to feed throughout the winter. By the following summer, the transformation to the adult form is complete.

Adult leopard frogs may remain in the swamps and marshes or may move to moist, grassy fields. While thriving on insects, worms, and small fish, they often fall prey to snakes, birds, larger frogs, and man.

Collected for food and laboratory use, the frogs are either shot or caught with baited hooks. As frogs' legs have become a more popular food item, population decline has generated protective laws in several states.

California Newt

Taricha torosa

DESCRIPTION

Classification: Amphibian

Size: Length to 8″

Characteristics: Warty, dark brown back; bright yellow-orange belly

HABITAT

Swamps, marshes

RANGE

Along the Pacific coast, primarily in California

AN afternoon drizzle in late December stirs the restless instincts of a California newt. From beneath the decaying remains of a fallen tree, the small glistening form slinks cautiously toward the water's edge. A young raccoon, sauntering toward his midday drink, detects the stirring amid the leaves and stops to investigate. Sensing the imminent danger, the newt commits himself to a previously successful ploy. As the raccoon approaches warily, the newt suddenly postures. Head thrust back toward an arching spine, tail curved forward over his body, and legs spread-eagle, the newt mimics a contortionist. The raccoon, however, unimpressed by these amphibious gymnastics, is startled by the flashing fiery orange warning of the newt's underside. Perhaps the raccoon's collective memory recalls the unpleasant taste of a previous encounter, and his attention returns to his thirst. Backing away, the raccoon then finds an alternate route to his drinking hole.

The California newt, his vent swollen with spermatophores, continues toward the standing water of the swamp. There an encounter with a female of his kind will initiate a brief courtship. Breeding among the California newts occurs between December and March. During this period the skin of the male becomes unusually smooth, presumably to enhance his proficiency in the water.

The newt's eggs are fertilized and released in large masses. The larvae hatch within three months displaying long gills and two balancers, tiny fingerlike projections that extend from each cheek. Within the same summer most of the larvae will lose their gills and leave the water as adults.

Western newts do not undergo the terrestrial eft stage common to the eastern newt. The adult newts remain in the water during breeding periods, but otherwise spend their lives on land. They generally burrow in organic debris quite near the water and are often active during damp weather.

Newts satisfy their hearty appetites with such succulent morsels as earthworms, insects, and other small invertebrates.

Eastern Newt

Notophthalmus viridescens

DESCRIPTION

Classification: Amphibian
Size: Length to 4″
Characteristics: Yellow belly; red spots in two rows along brown back; male with finned tail

HABITAT

Swamps, marshes

RANGE

Eastern half of the U.S.A.

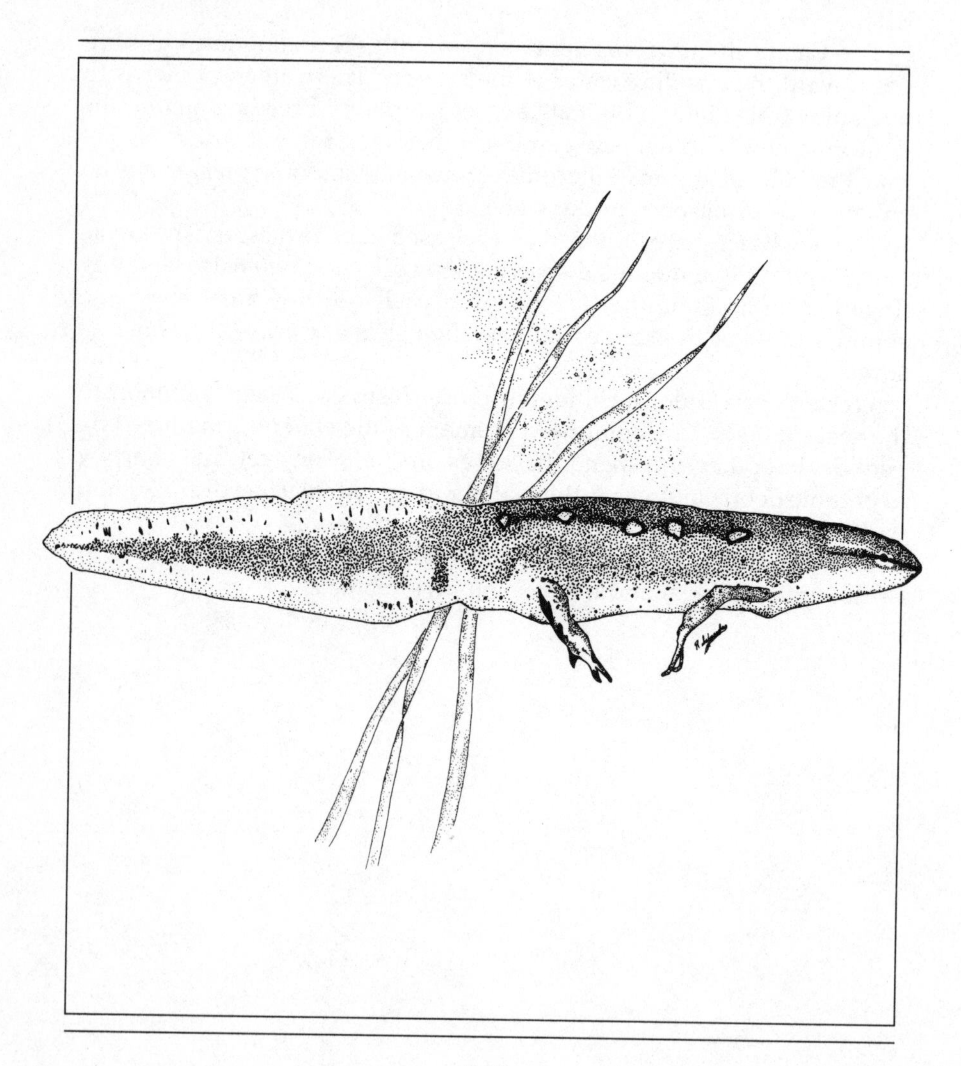

THE voracious appetite of the Eastern newt has been the ruin of many an ill-fated mosquito larva. Occasionally, a fish or amphibian youth adds gusto to the newt's monotonous but abundant diet of insects.

The aquatic adult, known to many as the red spotted newt, displays two rows of red spots along its brown back. The yellow underside is sprinkled with dark spots.

Employing a more sophisticated method of reproduction than frogs and toads, newts fertilize their eggs internally. During the courting process, the male deposits a gelatinous sperm-containing mass called a spermatophore. The female gathers the spermatophore into her cloacal chamber. As her eggs are about to be laid, they are fertilized. The eggs are then individually attached to submerged vegetation. The yellowish green larvae that emerge have external gills. By summer's end, the gills are absorbed and the emerging juvenile is a startling red creature richly spotted with black specks. The red eft is the terrestrial stage of the newt and is often associated with pixies and fairies. Living amid moist woodland debris, the eft slowly matures into an adult newt and returns to its aquatic beginnings.

In the Southeast, neotonic newts are often observed. In this condition, the terrestrial eft stage is eliminated and sexual maturity is reached without total transformation. Thus neotonic adults may have gill slits and/or reduced external gills throughout their adult stage.

Newts have a life span of seven years and may be kept quite successfully in aquaria.

Spring Peeper

Hyla crucifer

DESCRIPTION

Classification: Amphibian
Size: Length to 1½″
Characteristics: Tan to brown; X-shaped mark on the back with dark crossbands on the legs

HABITAT

Swamps, marshes, bogs

RANGE

Maine to northern Florida

THE spring peeper is a tiny frog heralding the spring. The shrill whistle emitted by the males is unmistakable and one of the first calls to be heard each spring. The intensity of the call is astonishing considering the size of the animal. A chorus of males can be heard as far away as one mile.

The color of the spring peeper is usually light coppery tan, with the males being somewhat darker than the females. The X-shaped cross on the back is the reason the frog was given the name *crucifer*. The color of the individual frog can vary daily depending upon its general activity. Environmental conditions such as light, moisture, and temperature also influence skin tone. The color of both sexes also changes to dark brown during the breeding season, providing better camouflage while mating. The animals have well-developed adhesive disks on their toes and are expert climbers, hence the name tree frogs.

In the spring, the males precede the females to the breeding area. Here their single throat patch under the chin inflates and they begin their characteristic whistle. The males also develop an exaggerated toe pad on each inner thumb. The whistling attracts the females, and mating occurs in the evening in shallow pools. The peeping reaches a crescendo on the warmest spring evening, but wanes or almost completely ceases if the temperature drops below 30° F.

After mating the female lays eight hundred to thirteen hundred cream and black eggs individually on underwater branches. It usually takes the female one day to complete this tedious task. The eggs hatch in five to fifteen days. Within three months the eggs develop into one-inch tadpoles with buff bellies and purplish blotches on their backs. In two more months the tadpoles assume adult form but will not sexually mature for three to four years.

Once breeding is complete, the peepers return to the trees where they mainly eat insects. They, in turn, are eaten by birds and squirrels. As a result of their small size and propensity for treetops, the spring peeper is not frequently encountered by the occasional naturalist.

Spotted Salamander

Ambystoma maculatum

DESCRIPTION

Classification: Amphibian
Size: Length to 7″
Characteristics: Wet, shiny, blue-black body with bright yellow spots

HABITAT

Woodlands to marshes

RANGE

Eastern half of the U.S.A.

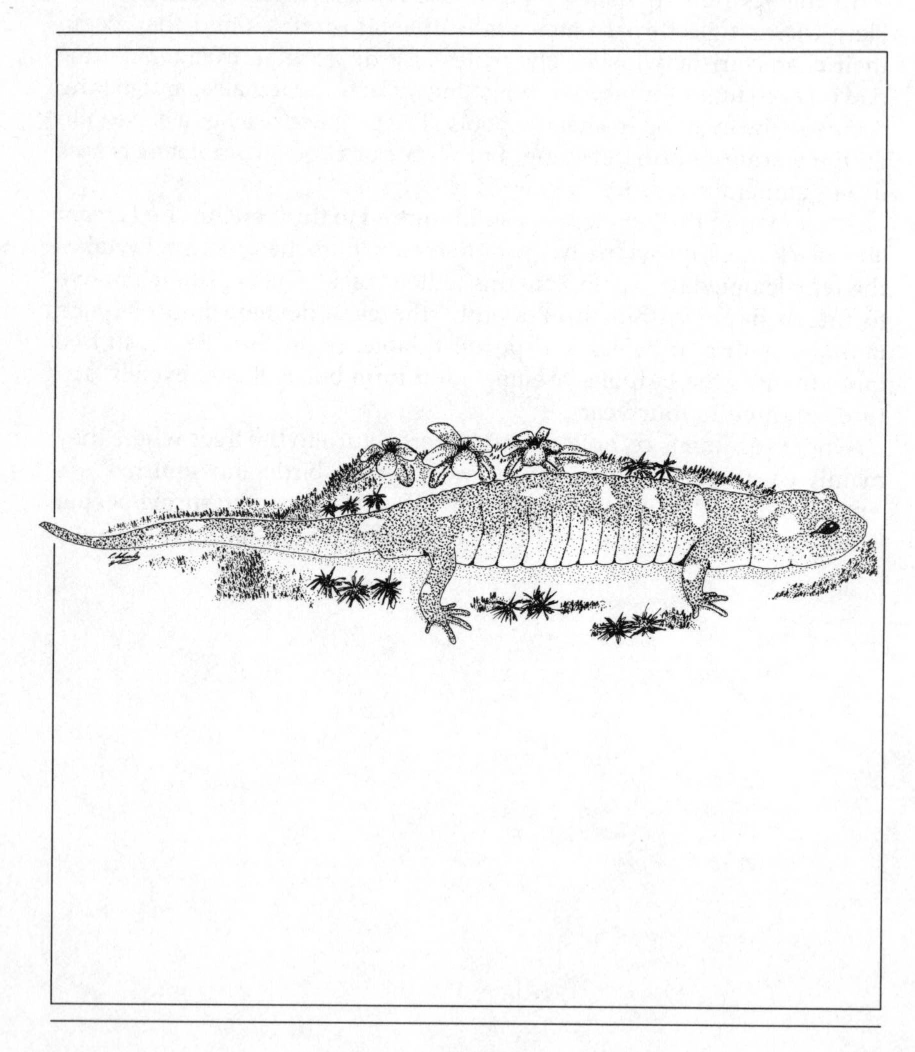

THE first mild rainy night in early spring prompts the mass migration of the spotted salamander. Glistening slate blue bodies obscured by patches of haze move in determined silence. Crossing paved roads or forest paths, salamander troupes seek shallow pools and marsh edges with instinctive resolve. Spurred onward by the primal urges of procreation, the spotted salamanders leave the comfort of their terrestrial habitat to rendezvous in the dark waters.

The courtship and mating habits follow a pattern typical of the entire order. The male discharges a jelly spermatophore, which is then assumed by the female through her cloacal opening to fertilize her eggs.

Nearly two hundred eggs may be laid in a single mass. As spring temperatures warm the water, tiny gilled larvae issue from this mass. By the dog days of August, the transformation to the adult stage accompanies the emergence of terrestrial habits.

Concealed among moist debris, spotted salamanders prey chiefly upon insects and slugs. When the salamanders are disturbed, a distasteful mucus is exuded from their body glands. Thus potential predators find them unpalatable. This effective feint has probably given rise to the many fictional accounts of their poisonous nature.

In recent years, salamander populations in the Northeast have become reduced. Habitat destruction and acid rain have been suggested as possible culprits. It appears that acid rain inhibits the proper development of salamander eggs. Fortunately, this problem is the subject of continuing study.

Oak Toad

Bufo quercicus

DESCRIPTION

Classification: Amphibian
Size: Length to 1¼″
Characteristics: Dark gray to green above; pale line extending along the center of the back from nose to tail; dark paired blotches on either side of center line

HABITAT

Swamps, marshes

RANGE

North Carolina to Florida; west to Louisiana

A CURSORY glimpse may suggest that Mother Nature has shortchanged the tiny oak toad in its struggle to survive. At barely over one inch in length, *Bufo quercicus* is among the smallest American toads. Its stubby legs and thick body limit its locomotion to short leaps, leaving it extremely vulnerable to predation. Its diminutive size, however, is used to its advantage as the toad can slip easily into tiny spaces or burrow quickly into leaves or soil. Most important in its encounters with adversaries is the whitish fluid emitted from the parotid and other skin glands. This material is so intolerably caustic to mucous membranes that most predatory animals need only one encounter in a lifetime to remain in awe of this tiny creature.

While touching toads is in no way related to developing warts, other unpleasant consequences may result. Those of us who insist upon handling these ostensibly innocuous amphibians should be advised that residues of their secretions may remain upon our hands. Only thorough washing will assure that unnecessary irritation of the eyes and mouth will be avoided.

Breeding choruses of oak toads inspire the same incredulity as those of the northern spring peeper. That such a resounding volume of chirp-like calls can be the voice of such a tiny creature is not less than amazing. While the oak toad is mainly a land dweller, his amphibious roots dictate his return to the water for mating.

The long strands of eggs laid in the water quickly develop into a profusion of tadpoles, which will furnish nourishment for many fish. Those tadpoles that succeed in becoming terrestrial toadlets do so with an accompanying thickening of their epidermis into a heavy warty skin. Masses of toads then swarm out of the water and begin the terrestrial phase of their lives.

American Alligator

Alligator mississippiensis

DESCRIPTION

Classification: Reptile
Size: Length to 20′
Characteristics: Dark, leathery, rough skin with pale crossbands

HABITAT

Swamps, marshes

RANGE

North Carolina south to Florida; west to Texas

WHEN Spanish sailors first encountered alligators in the rivers of Central America, they called the giant reptiles *el largota,* meaning "the lizard." Later as the English explored the New World, *el largota* was Anglicized to "alligator."

To the casual observer the alligator's broad, blunt snout easily distinguishes it from its crocodilian cousin. Further investigation reveals extensive differences in the dental structure, with the alligator having an obvious overbite.

Alligators are well known for their long, sluggish lives. Many have been documented to be over fifty years old. It has been suggested that their torpid nature contributes to their longevity. Inhabiting areas below the thirty-fifth latitude, alligators are intolerant of cold weather. They commonly hibernate during the coolest weeks of each year.

Upon awakening in the spring, male alligators initiate breeding with bellowing roars and brawling with other males. These actions and the musky emissions of the males attract the females to the breeding site. Mating among the alligators is a rather tumultuous process that is usually performed in the water after dark.

Females build large mounds of mud and organic debris. Dozens of eggs (fifteen to eighty) are laid on top of the mound and then covered with more moist debris. For two to three months the clutch is incubated with heat from the decaying vegetation. During this time the female will remain in the general area of the nest to protect it. Loud peeping from the hatchlings signals the female to uncover the nest and welcome her young to the world. Mother and young may remain together for as many as three years.

Insects and freshwater shrimp provide the hatchlings with nourishment. As the hunting skills of the young become more adept, fish, frogs, and snakes will comprise the bulk of their diet. Adult alligators will also prey upon small mammals and waterfowl. An average mature alligator can swallow a duck whole. Hatchlings and juveniles often fall prey to fish, birds, mammals, and even other alligators.

The American alligator population has been the product of nature's own balance for centuries. Three aspects of our culture have greatly reduced this population, even threatening extinction within our lifetime. Hunting for skins, the pet trade, and land drainage have had a devastating effect on alligator numbers. Fortunately, state and federal regulation has allowed the alligator to make a rapid comeback. It has recently been removed from the endangered species list of Louisiana, Texas, and Florida and now enjoys a special status. Current law permits limited hunting while also granting the state government the right to quickly reimpose tight controls if necessary.

Alligators have a somewhat legendary status in New York City. At the height of the alligator pet trade, many lively young alligators were purported to have been disposed of in the city's sewers. Despite official denials, there have been many unsubstantiated reports of sewer-raised alligators preying upon rats and sewer workers beneath the city streets.

Bog Turtle

Clemmys muhlengergii

DESCRIPTION

Classification: Reptile
Size: Length to 4″
Characteristics: Striking orange patch on neck

HABITAT

Marshes, bogs

RANGE

Rhode Island; Connecticut; New York; New Jersey; Pennsylvania; Maryland; Virginia; North Carolina (isolated colonies)

PEERING cautiously from beneath a tussock sedge, the grasses draped about its neck and face, a bog turtle ventures from its overnight burrow. The bright orange splash on the side of its extended neck appears to have been dabbed on with a paintbrush. The bog turtle will spend the next few hours basking on the tussock. Revitalized by the sun's warmth, the turtle then embarks on the elemental task of foraging.

For many years it was commonly accepted that bog turtle populations were dangerously declining. The federal government has the bog turtle listed as a rare and endangered species. Many states, as well, offer the turtle protection from amateur collection and, to some degree, from habitat destruction. More extensive research now suggests that perhaps the bog turtle is a wary, even somewhat crafty reptile, who is less then enchanted with human encounters. Combined with its small size (it is among the smallest of the North American turtles), the discreet bog turtle easily avoids the most vigilant wetlands visitors.

Bog turtles are generally active during daylight hours. They enjoy sunning except during very hot weather, when they seek the shelter of their burrows. Their omnivorous feeding habits lend great variety to their diets. Berries, seeds, and other vegetation are complemented with tadpoles, slugs, snails, and insects. Unlike some other semiaquatic turtles, bog turtles are capable of swallowing their food while remaining both in and out of the water.

Reaching sexual maturity in five to seven years, bog turtles seek the secluded waters of swamps and bogs for breeding. They mate each spring, nesting in June. The eggs are elliptical, numbering from one to six in a clutch. The hatchlings, emerging in late summer, are quite circumspect and are more aquatic than the adults. The dark brown carapace is quite rough in young turtles but becomes smoother with age. Females sport a shorter, thinner tail than the males, who have a vaguely concave plastron.

Snapping Turtle

Chelydra serpentina

DESCRIPTION

Classification: Reptile
Size: Length to 3′; females larger than males
Characteristics: Ridged, rough carapace with serrated rear edge

HABITAT

Swamps, marshes, bogs

RANGE

Eastern two-thirds of the U.S.A.

THE snapping turtle is an animal of superlatives. It is the largest, most abundant, and most maligned North American turtle. Often exceeding thirty pounds, braced in its aggressively offensive posture with powerful jaws agape, it can strike fear in a super hero. The large head and jaws effect a powerful bite capable of inflicting serious injury.

The body structure of the snapping turtle, particularly the shell, is reminiscent of some long-extinct species. Often stained with mud and veiled in algae, the dark, rough carapace is both ridged and serrated. The unusually long tail is also toothed along the upper midline. Knobby protuberances on the neck and thick scaling on the legs and face add to its grotesque appearance.

Snapping turtles are actually out of their element when on land. This probably accounts for their vicious response to threat, which has gained them notoriety. They tend to be quite diffident in the water, however. Camouflaged by mud and foliage, often with no more than eyes and nostrils exposed, snapping turtles are more likely to retreat from confrontation than to initiate it.

They have voracious appetites which are satisfied with vegetation, animals, and carrion. Snapping turtles are widely stigmatized as controllers of the duckling population.

Snappers will mate during any season except winter. Six to seven dozen eggs may be found in a single nest which is usually a considerable distance from water. Incubation time varies with the climate. In the northern habitats hatchlings often remain in the nest over the winter.

Snapping turtles are widely used in soups and stews. Well before Europeans ventured into North America, the Indians regularly included snapping turtles in their diet.

Southern Soft-Shelled Turtle

Trionyx ferox

DESCRIPTION

Classification: Reptile
Size: Male to 12″; female to 20″
Characteristics: Round, flat, leathery brown carapace; very long neck; yellow or red stripe darkly outlined connecting each eye and lower jaw

HABITAT

Swamps, marshes

RANGE

South Carolina; Georgia; Florida; Alabama

THE soft-shelled turtle is probably the most bizarre of all the chelonians. Quite unlike the more common members of the turtle order, it is very well adapted to its habitat and lifestyle. The rigid, domed calcium-rich shell normally associated with turtledom is replaced with a flattened overlay of tough skin. The perimeter of the so-called soft shell is completely boneless, leaving it thin and supple. This type of shell is a more streamlined structure and is rather efficient for water travel.

Soft-shelled turtles are, in fact, quite agile in the water. Fully webbed feet and strong musculature make them more than competent swimmers. Soft-shells so prefer their aquatic environment that even reduced water levels rarely press them ashore. During droughts the turtles have often been seen burrowing deeper into the mud rather than seeking water elsewhere.

Another striking feature of the southern soft-shell is the long, narrow snout. Coupled with its very long, flexible neck the animal easily remains concealed underwater with just its nostrils breaking the water's surface. Almost comical in appearance, these turtles pass many hours each day suspended in the water or resting on the muddy bottoms of shallow pools.

The fleshy mouth conceals powerful jaws and a sharp, horny beak. This, along with sharp claws and a nasty disposition, makes it a formidable adversary for prey and turtle tormentor alike. These adaptations, both physical and behavioral, allow the soft-shelled turtles to successfully meet the demands of a carnivorous lifestyle. Their favorite fare includes crayfish, frogs, fish, and even young waterfowl.

Each spring the nesting instinct drives the female southern soft-shell to land. Usually fewer than twenty eggs are laid in the shallow depression dug in the moist margins. The spherical eggs are about one inch in diameter and have rather thin shells. A nine-week incubation period follows, although the time may vary with soil temperatures.

In its native regions the soft-shelled turtle is sought for its food value. It is considered far tastier than the snapping turtle and is marketed commercially.

Black Swamp Snake

Seminatrix pygaea

DESCRIPTION

Classification: Reptile

Size: Length to 20″

Characteristics: Dull violet-black above; scarlet underside

HABITAT

Swamps, marshes

RANGE

Coastal plain of North Carolina, South Carolina; southern Georgia and Florida

AS the morning sun creeps above the horizon, caution bows to sentience and many wetlands reptiles crawl from their nightly cover in search of the sun's warming rays. For animals like the black swamp snake this morning ritual is concluded with several hours of sunning. But unlike its more aggressive cousins, *S. pygaea* is more likely to remain at least partially camouflaged with wet leaves, sphagnum, or pine needles. Once its body temperature is high enough to resume its daytime activities, the swamp snake slips deeper into the safety of aquatic vegetation. Especially fond of water hyacinth mats, *S. pygaea* may hide within the roots for extended periods.

This species is rather timid and reportedly very easy to handle. Apparently enjoying a quite varied diet, it has been observed consuming worms, leeches, small fish, tadpoles, frogs, and salamanders. Viewed from above, the black swamp snake is an inconspicuous reptile, not particularly long and certainly not brightly colored. But this masked deception fails when the animal moves, revealing glimpses of brilliant red along its lower surface. The three related subspecies of *S. pygaea* are commonly referred to as red-bellied mud snakes.

S. pygaea makes it first springtime appearance between February and March. For the females, approximately five months of gestation will follow the spring mating. By fall, the gravid females will retire to their concealed nests, where they will bear small litters of live young. By December, most individuals will again be nestled in their winter hibernation sites.

Water Moccasin

Ancistrodon piscivorus

DESCRIPTION

Classification: Reptile
Size: Length to 6′
Characteristics: Olive, brown, or black above with dark crossbands becoming less distinct as the animal ages; lighter belly; interior of mouth is white

HABITAT

Swamps, marshes

RANGE

Southeastern Virginia to Florida; west to Alabama

AS the noonday sun's rays flood the southern swamp, the humidity becomes oppressive, and a pronounced stillness descends over the area. The wildlife retreat quietly to the shadows to await the evening respite. The water moccasin welcomes this tropical fever and settles cozily into the forked branches of a nearby shrub. Basking in the sun's warmth, the snake's metabolic rate quickens, aiding in the digestion of its most recent meal. Preferring to feed at night, the water moccasin, also known as the cottonmouth, consumes mainly amphibians and other reptiles. Fish, birds, and small mammals, however, are not exempt from the cottonmouth's diet.

The cottonmouth is fairly adept at avoiding such predators as birds, mammals, and other reptiles. As the weather cools, these enemies pose an even greater threat owing to the snake's more sluggish movements. Snake collectors exploit this reptilian characteristic, but are held in check by federal law.

Mating and gestation habits in the wild vary greatly from those in captivity. In their natural habitat, breeding activities occur in the spring. The young, strikingly marked with bright yellow tails, are born live between August 15 and September 15. In captive habitats, where light and temperature are uniformly maintained year round, matings have been reported throughout the year. Vigorous courtship dances precede mating. Inasmuch as the female still gives birth from August through September, some questions are being raised about the correct gestation time.

Numerous reports have been documented of encounters with water moccasins in the wild. The snakes do not seem to demonstrate an aggressive nature if mildly disturbed. When captured and handled, however, they become belligerent and gape widely to display the pale interior of their mouths. This characteristic is the source of the common name of cottonmouth.

Although virulent, the venom of the water moccasin is rarely deadly. The hemotoxic nature of the fluid destroys the red blood cells of the victim and coagulates blood at the wound site. The venom is less potent than that of copperheads and eastern diamondback rattlesnakes and has been used medicinally in the treatment of both severe hemorrhage and rheumatoid arthritis.

Northern Water Snake

Natrix sipedon

DESCRIPTION

Classification: Reptile
Size: Length to 50″
Characteristics: Dark brown to black above with light crossbands; underside marked with alternating, vaguely triangular, black or yellow pattern

HABITAT

Swamps, marshes, bogs

RANGE

Eastern half of the U.S.A.; Kansas, Nebraska, Oklahoma

THE northern water snake, dull in color and sluggish in habit, is frequently stumbled upon by the serious naturalist. When seen at eye level draped on an overhanging branch, a four-foot snake with a girth of four inches and a sinister expression can be quite startling. Coupled with the general bad press suffered by snakes, it can even be alarming.

Although nonvenomous, the water snake is an aggressive and formidable creature capable of inflicting a painful wound. When aroused it will usually strike viciously and repeatedly, injecting its antagonist with an anticoagulant. This can leave the wound site aching and profusely bleeding. In spite of this behavior, the water snake can be easily tamed, thus often falling victim to amateur collectors.

Natrix sipedon and its related species are commonly encountered throughout the country. They are abundant in both numbers of species and numbers of individuals. A large specimen of *N. sipedon* gliding through the water can easily be mistaken for the venomous water moccasin.

The monotonous color of the thick-bodied adults creates no illusion of beauty. The dark, dull skin only takes on a gleaming brilliance for a short time following each periodic shed. For nearly a week before each shed the skin tones are further obscured by a cloudy haze effected by the loosening outer layer of skin tissue.

Individuals of *N. sipedon* mate shortly upon awakening from their winter rest. Most are actively feeding, sunning, and mating (often in masses) by late February or early March. Throughout the spring and summer months, appetites are satisfied with fish, amphibians, crayfish, and small rodents, consumed whole and alive. By late August or September females nest to bear live young. Up to seventy snakelets have been reported in a single litter, but fewer than thirty is more common. Within days the young abandon the nest site to seek food and fend for themselves.

By early October the water snake once again seeks the shelter of its hibernation site, either in a rodent burrow or the crevice of a rock.

Anhinga

Anhinga anhinga

DESCRIPTION

Classification: Bird
Size: Length to 36″; wingspan to 47″
Characteristics: Long, slender neck; pointed bill; long tail; silvery wing patches; *male:* green-black and glossy; *female:* tawny neck and breast; *immature:* brownish

HABITAT

Swamps, marshes

RANGE

Atlantic and Gulf Coasts from North Carolina to Texas; north to Arkansas and Tennessee

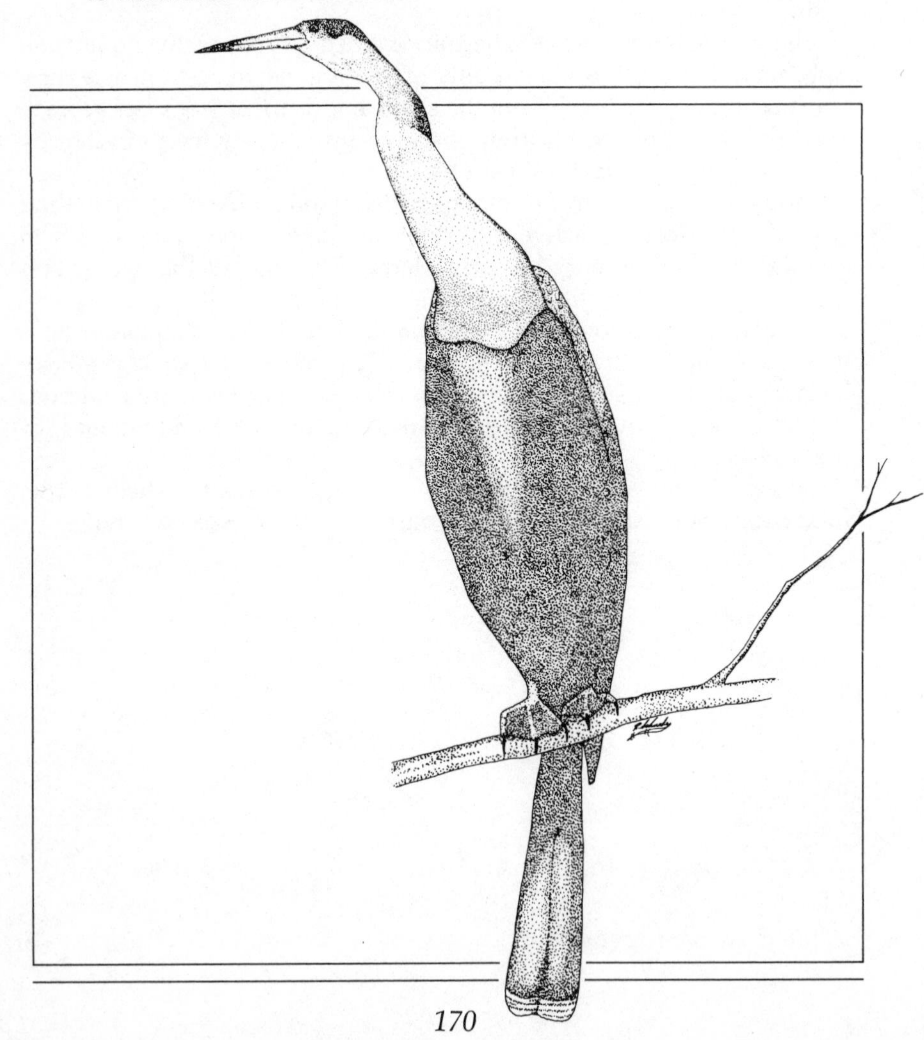

SILHOUETTED against the rosy pink light of dawn, the tangled swamp appears to be forsaken. Slowly, the faint stirrings of its avian residents begin. As the misty light grows in intensity, so does the cacophony of voices, suggesting a joyous greeting of the new day. A large, almost prehistoric-looking bird, however, does not join in this revelry but perches silently on a dead branch overlooking the still water. Suddenly startled, this long-necked anhinga drops like an arrow into the water creating only a faint ripple. Reappearing moments later with only its slender snakelike neck in evidence above the surface, the anhinga moves slowly away.

The anhinga, also known as the darter, is a slender relative of the cormorant, living almost exclusively in freshwater areas. The male is an iridescent green-black bird with a long, slender neck, thick plumage, a long tail, and silvery wing patches. The female resembles the male except for her buff-colored head, neck, and breast, while the immature bird is completely brown in color.

Enjoying a diet of frogs' eggs, insects, fish, and an occasional small alligator, the anhinga is an adept fishing bird. Unlike a heron, which stalks the fish from above the water, the anhinga dives from the surface of the water and swims underneath in the pursuit of its meal. The anhinga stabs at its prey with its beak open, thus doubling its chance of success. Upon impaling the catch, the bird must flip the fish into the air to turn it around. This allows the anhinga to swallow the fish headfirst and whole.

Swimming with body submerged and only head and neck protruding from the water, the anhinga is often referred to as the snakebird. Lacking oil glands with which to preen its feathers, the bird must balance spread-eagle on a branch in order to dry its water-laden wings. Appearing ungainly while perched, the anhinga is a strong and graceful bird in flight. Neck extended and long tail spread wide, it alternately flaps and glides and can soar to several hundred feet on motionless wings.

Anhinga often nest among herons and ibises. Initially the male claims a nest site and then advertises for a female by wing waving and bowing. After a female selects a mate and his site, and is accepted by the male, she begins to build a nest with materials brought by her mate. Usually composed of twigs and Spanish moss and lined with soft, green leaves, the nest is placed in the fork of a tree anywhere from five to one hundred feet above the ground. After mating, two to five chalky blue eggs are laid. Upon hatching the chicks are fed regurgitated food by both parents. The chicks will be protected until they are old enough to defend themselves.

The anhinga, by nature, is a shy and retiring bird, and although it is usually silent, it is capable of emitting soft, low grunts. Found in deep wilderness, the anhinga's life seems to be pervaded with mystery. Since habitat destruction is causing the number of anhinga to decline, the fate of this elusive bird remains in question.

Red-Winged Blackbird

Agelaius phoeniceus

DESCRIPTION

Classification: Bird
Size: Length to 9½″
Characteristics: *Adult male:* black with red and yellow shoulder patch; *immature male:* mottled brown with red shoulder patch; *female:* dark brown, heavily streaked

HABITAT

Swamps, marshes

RANGE

Continental U.S.A.; breeding range extends throughout continental U.S.A.

IN early March, large flocks of male red-winged blackbirds, advancing on a wide front and wheeling in unison, can be observed heading north to places where they have nested in the past. Their song, a pleasant *kon-ke-ree,* seems to proclaim a triumph over winter and is considered one more harbinger of spring.

As this legion of ebony males maneuvers overhead in an undulating flight, rising gently on beating wings and dropping sharply on glides, their buff-bordered, scarlet shoulder patches flash brightly. Arriving in a chosen area, the group will scatter slightly, then begin the process of singing and displaying as they compete for breeding territory. The males generally arrive two to five weeks prior to the females and can live quite close to one another, as long as they honor each other's territorial rights.

The dusky brown females have heavily streaked breasts and are slightly smaller than the males. Upon arrival, several females will vie persistently for the attention of one male and will compete among themselves for a portion of his territory.

After mating, coarse grasses are woven into a bulky, bowl-shaped nest, which is then lined with soft, fine grasses and rootlets. The nest is attached to reeds or shrubs at heights of less than fifteen feet above the ground. Three to five pale, bluish green eggs marked with irregular brown, purple, or black streaks are laid and incubated for ten to fourteen days by the female. Each pair usually raises two to three broods a year, building a new nest for each clutch.

Red-winged blackbirds gather in extensive flocks, called "ranks," in the fall, winter, and spring. During the summer, however, the birds hide in the vegetation in order to molt their flight feathers and grow new ones. The flocks then reassemble in late August to migrate south.

Feeding mainly on marsh plant seeds, including those of the cattail and wild rice, red-winged blackbirds will also eat grain, fruit, and insects in season. Cankerworms offer a delightful morsel for the blackbirds. During the spring breeding period the fields and sky will be teeming with birds capturing food for their young. By summer's end, however, the ripened corn fields sorely tempt the ranks of red-winged blackbirds. Resorting to local raids, the birds effect extensive damage to the crops.

Since red-winged blackbirds are gregarious, they congregate by the thousands. Trees swarming with roosting birds are a common sight and one that poses a complex problem. Besides the steady din produced by the chattering multitudes, the proximity of this huge mass of birds creates a health hazard both to the birds and to man. Government regulation of these swarms is usually accomplished by destroying the birds. When some of the birds are killed, the survival rate among the young born the next year increases, and in one year, the population is restored to its previous level. If all of the birds are destroyed, however, the insect problem during the following spring and summer becomes intolerable.

Generally the beneficial aspects produced by the red-winged blackbirds outweighs the negative aspects, and we must find a way to balance our interests with theirs.

Wood Duck

Aix sponsa

DESCRIPTION

Classification: Bird
Size: Length to 20″; wingspan to 29″
Characteristics: Male: rainbow iridescent body, swept-back crest; white chin patch, long tail; *female:* dark brown body, white eye ring

HABITAT

Swamps, marshes

RANGE

Continental U.S.A.

LIKE shining jewels on a necklace, the brightly colored iridescent wood ducks often line up in a row on a dead tree branch perched over the water. Basking in the sun, some sleeping, some preening, they appear contented with their lives.

The male wood duck is one of the most prismatic birds of the United States with its pattern of green, purple, yellow, red, and blue. The female, somewhat smaller and more darkly colored than the male, sports a white eye ring. This handsome bird is more often heard than seen, and its voice consists of a loud *whoo-eek*, with softer *peet* and *cheep* notes.

Agilely dodging trees, the wood duck is a rapid and direct flyer, holding its head above body level with bill pointed downward. The wood duck exhibits grace while swimming and diving as well as the characteristic waddle while walking. Both traits are due to the exaggerated rear positioning of the bird's legs.

The wood duck feeds mainly on duckweed, tubers, galls, cones, seeds, insects, and acorns with the young occasionally snacking on mosquito wrigglers. Arriving after harvest and taking spilled grain, wood ducks are often accused of damaging farm crops, although there is little evidence to support this claim. Wood ducks do, however, especially enjoy the acorn of pin oak trees and will seek these out each fall.

Although the wood duck is a migratory bird, it breeds over nearly all of the United States. Returning to the same nesting area year after year, the female brings with her the drake who paired with her in midwinter. Here she seeks out a hollow tree to nest. Since the removal of old timber has greatly reduced suitable nest sites, nesting boxes have been substituted and appear to be working successfully.

The female lines the nest, which may be placed from three to fifty feet above the ground, with gray down plucked from her breast and lays from eight to fifteen pale brown eggs. Incubation takes from twenty-eight to thirty days and is accomplished mainly by the female with the drake present less and less. Slightly before hatching time, the drake deserts his mate and leaves the care of the young to her. Soon after hatching, the ducklings jump from the nest hole in response to their mother's call and are led to water. They will remain with the female until the end of the season.

Upon leaving the female and her hatchlings, the adult males rendezvous to produce flocks of twenty to thirty. They disappear from their breeding areas and enter secluded water habitats. Here they molt, losing all of their flight feathers simultaneously, and are unable to fly for weeks. By the end of July, this molt is complete and they have acquired the eclipse plumage. In this eclipse stage, the adult male closely resembles the immature male. By the beginning of September, most males molt again and acquire their complete winter or breeding plumage for the next nesting season.

Females acquire their adult plumage after losing their down and also undergo a yearly molt in August or September, when they, too, lose all of their flight feathers. The females, however, do not undergo the double summer molt and the eclipse plumage.

Suggested as a national bird because of its peaceful habits and beauty, the wood duck nests in virtually every state. The large size of the wood duck, however, has made it the quarry of man and its period of flightlessness has been exploited. As populations dwindled the species was protected by federal law. Today, despite increases due to the protection and artificial breeding of wood ducks, the population continues to suffer at the hands of man.

Purple Gallinule

Porphyrula martinica

DESCRIPTION

Classification: Bird

Size: Length to 14″; wingspan to 23″

Characteristics: Head and underparts deep purple; green back; pale blue forehead shield; red bill tipped with yellow; yellow legs; *immature:* dark above, pale below; dark bill

HABITAT

Swamps, marshes

RANGE

Carolinas and Tennessee to Florida and Texas; occasionally wanders north to New England

STRIDING confidently over the floating lily pads with a continuous bobbing of its head, the purple gallinule seems to take pleasure in displaying its brightly colored plumage. With its deep violet-purple head and underparts, its pale cobalt blue frontal shield, its bronze green back, its chrome yellow legs, and its red and yellow bill, the gallinule is well camouflaged in its home of blue water, green leaves, and multihued flowers.

The purple gallinule is a long-necked, long-legged bird with a body that is laterally compressed. These traits allow it to move quickly, quietly, and easily through the intricately meshed plants of the marsh, since it prefers to run rather than fly. With its long toes, the gallinule can walk quite expertly upon floating vegetation, clamber rapidly about shrubs and trees, and climb vines with a flair. It can also swim well, even though its feet are not webbed.

With short, rounded wings, the gallinule exhibits a feeble and hesitating flight. Feet dangling, the gallinule emits a henlike cackling as it flies weakly over the wetlands. Despite this display, however, the bird is a migratory species capable of flying hundreds of miles to winter in South America.

The purple gallinule, with its short, heavy bill, is well adapted for feeding upon frogs, shellfish, aquatic insects, seeds, and flowers. Quite a hunter, the gallinule consumes the majority of its food under dense vegetative cover.

Appearing in its summer home by April or May, the purple gallinule enters the courtship phase. After mating, nest building begins in the reeds over the water. Varying from two to five feet above the water level, the nest, constructed of dead grass and rush stems, is completely hidden from view by surrounding vegetation. Six to ten creamy white eggs, spotted with brown and lavender, are laid in a single clutch. Newly hatched chicks are covered with black down and have white-tipped hairlike feathers on their head. The young chicks can leap from the nest and run about soon after emerging from their shells. Before obtaining their adult plumage, the immature gallinules are dark brown above and pale below.

The purple gallinule is considered a game bird, but its numbers are not extensive. The bird does no harm to man's interests and is more valuable alive in a marsh than dead as a hunter's catch.

Marsh Hawk

Circus cyaneus

DESCRIPTION

Classification: Bird
Size: Length to 24″; wingspan to 42″
Characteristics: Long wings and tail, white rump; *Male:* pale gray with black wingtips; *female and immature:* brown above, streaked below

HABITAT

Marshes

RANGE

Continental U.S.A.; breeding range extends northward from Pennsylvania

FALL is a time of constant activity in the wetlands. Crowded with birds in transit, the vegetation strains under the weight of those alighting to feed, drink, and rest. Marsh hawks glide low over the area with their wings angled upward. Tilting from side to side, the marsh hawk methodically scans the terrain with the care and precision of a well-trained hunting dog. Suddenly its forward flight is abruptly arrested, and it drops sharply downward, talons outstretched, onto its surprised victim. The marsh hawk is a relentless hunter, as it pursues or "harries" its rodent prey, and when caught the victim is devoured on the spot.

Raptors, or birds of prey, have captured the imagination of people throughout history. Not only are the hooked beak, penetrating eyes, and strong talons indications of great power, but an aura of mystery has surrounded these creatures since they are very rarely seen by man.

The marsh hawk, or northern harrier, as it is often called, is the only representative of the harrier group in North America. The females are dusky brown above with brown streaked undersides, while the males are gray with black wingtips. Both sexes have a white rump, and a black and buff, barred tail. Their call is a *kee-kee-kee* or sharp whistle.

Marsh hawks migrate yearly, beginning in August, to the more southern reaches of the United States, where they will overwinter. Hawks in flight can be observed migrating along lake and ocean shores as well as mountain ridges which are excellent sources of updrafts. The hawks are specially adapted to make use of the convective air currents to help them remain aloft for long periods of time with minimal wing movement. This form of energy conservation enables them to travel hundreds of miles a day without having to consume extra food to support their energy requirements.

The courting maneuvers of the male involve soaring high in circles on level wings, then dropping to perform adeptly executed somersaults in midair. They are done in complete silence; only the sound of rushing air along the wings can be heard.

After mating, both partners help to construct the ground nest among the vegetation. Composed of grasses and reeds, the same nest is used year after year, and the structure eventually gets bulky and resembles a platform. Four to six bluish eggs are laid and are incubated by both parents for twenty-six to thirty-one days. During this time, the raptors are not easily observed because they remain perched on the nest without stirring. Upon hatching, the nestlings are blind and quite helpless. The protective parents become quite aggressive in the face of unwelcome visitors. Diving threateningly and shrieking wildly, the hawks effectively frighten off intruders. The young are fully fledged within four months but remain with their parents for nearly a year until they grow independent. By three years most young will have exchanged their juvenile appearance for adult plumage.

Marsh hawks feed primarily on small mammals and birds but will also eat insects, snakes, amphibians, and occasionally poultry. It has been claimed that the marsh hawk is responsible for heavy losses to poultry; however, evidence has proved this to be false. Since the hawk feeds primarily on the rodents that destroy agricultural crops such as corn and cotton, this raptor is of vital importance in keeping the populations of these prolific breeders in check.

The marsh hawk is protected by law in many states and is very useful to man in controlling agricultural pests. The marsh hawk is also used as an environmental barometer. Perched ecologically at the top of the food web, it is vulnerable to environmental damage and one of the first to succumb to poisons. Unfortunately, habitat destruction, pesticides, and wanton killing by man are reducing its numbers.

Great Blue Heron

Ardea herodias

DESCRIPTION

Classification: Bird
Size: Height to 50″; wingspan to 70″
Characteristics: Slim, slate gray body with long neck; long, brown legs; yellow bill; flies with neck folded; *adult:* white head with two black plumes; *immature:* black crown with no plumes

HABITAT

Swamps, marshes

RANGE

Continental U.S.A.

STANDING motionless amid the tall marsh reeds, its long neck extended upward, the great blue heron maintains an alert vigil over the nearby water. As a fish approaches, the heron holds his stance until the creature is within striking range. With a lightninglike thrust of its neck, the heron seizes the hapless prey. Adroitly the fish is turned and swallowed headfirst and whole. The heron then reassumes his rigid pose. Herons are in constant search of food to satisfy their high caloric needs. Often observed stalking their prey, herons pursue fish, frogs, crustaceans, snakes, mice, and birds. Lifting each foot carefully, so as not to disturb the water, the heron stealthily approaches its unwary victim, seizes it, and feeds again.

The great blue heron is a large, gray, cranelike bird. With its long legs and long toes adapted for walking on mud, it is well equipped for wading. The young resemble the adults in coloration, except for their black crowns and lack of plume feathers. The great white heron, found in Florida, is actually an all-white variety of the great blue heron and not a separate species as once believed.

Appearing to be quite bulky when floating, the heron can take off quickly and effortlessly from the water. Flying with its neck in an S-curve and its long legs extended behind its tail, the silhouette in the sky is unmistakable. Deliberate, powerful, and certain, the heron maintains a slow, steady flight with heavy wing beats and can attain a speed of approximately twenty-eight miles per hour.

Mainly solitary creatures, herons become gregarious during the breeding season. It is then that they form large colonies, generally in isolated swamps. Arriving in small groups by the end of February, they gather at the base of the nesting trees in the heronry. The adults remain in this "standing area" all facing in the same direction. As the group enlarges, a synchronization of breeding cycles is achieved, and it is at this point that the herons begin to occupy the treetops. The male advertises for a female by posturing, and mating soon follows. A platform nest of twigs, lined with soft materials, is built. Three to five pale, dull blue eggs are laid and are incubated by both parents for twenty-six days. The young are helpless and are fed at first by violent regurgitation by the parents. During the spring, the heron rookery may contain dozens of nests and be bustling with activity.

The economic importance of the great blue heron is as yet undetermined, but it is protected by law in many states. With continued perseverance this stately and beautiful bird will be enjoyed by future generations.

Barred Owl

Strix varia

DESCRIPTION

Classification: Bird
Size: Length to 24″; wingspan to 45″; weight to 2 lbs.
Characteristics: Round head; stocky body; brown eyes; mottled brown with dark crossbands across chest and vertical streaks on belly

HABITAT

Swamps

RANGE

East of Rockies

ON a nearly moonless night in the secluded swamp, the sliver of gold hanging in the sky is periodically obscured by fleet-footed clouds racing overhead. The soft scurry of a field mouse, faintly audible to humans, alerts the sensitive ears of a vigilant nocturnal predator. Suddenly a large, dark form wings silently past and swoops down unerringly upon its tiny target. A quick closing of the deadly talons, and the mouse is silenced forever. The barred owl takes flight with its evening quarry dangling in its grip. After finding a perch upon which to feast, the owl's appetite will be briefly satiated.

The barred owl prefers the isolation of quiet, unfrequented swamps. It is here that this dynamic denizen reigns supreme, challenging man's presence with its doglike barking call. On certain nights the continuous serenade of the barred owl can be heard as a rhythmic cadence. Consisting of eight quick hoots in succession, with five to seven series per minute, the call is distinctive. Occasionally, however, a blood-curdling scream is emitted that sends chills down one's spine. The barred owl is a very inquisitive creature and will often reply to a human imitation of its call, even one somewhat crudely attempted.

The gentle appearance of the barred owl with its soft plumage, barrel-shaped body, and round baby face contradicts the true fierceness of this powerful predator. Equipped with a sharp, hooked beak and strong talons, it is a highly skilled hunter. The barred owl tends to be quite possessive of its territory and will attack intruders throughout the year.

The mottled brown color of the barred owl is similar in both sexes, but the female is usually larger in size. Horizontal crossbars found on the throat and vertical streaks found on the chest are the distinguishing characteristics of this species. The large, brown eyes have excellent depth perception and binocular vision. Lined with light-sensitive cells, the eyes are well adapted to night vision, and even the dim light of a candle burning one thousand feet away is enough to allow the owl to catch a mouse. Since the eyes are fixed in their sockets, the owl must rotate its head in order to shift focus.

The facial disk of the owl conceals large external ear flaps. The ears, which are asymmetrically located, admit slightly different sounds to each of the owl's eardrums. This unique perception of sound enables an owl to pinpoint the location of its prey. The owl has specialized wing feathers that enable it to fly silently. Since the edges of the flight feathers are soft, no sound of rushing air can be heard as the owl flaps its wings.

The barred owl begins nesting as early as March and often claims a hollow tree or a deserted hawk's nest as its home. After mating the female lays two to four glossy white, rough eggs, which she incubates for twenty-eight days. The young are white and downy when hatched. Only one annual brood is raised per nesting pair.

Consuming 10 to 12 percent of its body weight per day in food, the barred owl prefers mice but also enjoys frogs, lizards, crayfish, spiders, insects, and small birds. Occasionally, though, a domestic chicken that is forced to roost in a tree at night becomes fair game. The preference of the barred owl for a diet of mice is an indication of its economic benefit to man. Unfortunately the barred owl is not always appreciated. This elusive bird is protected by law in many states as its numbers are declining due to habitat loss.

There are three subspecies of the barred owl found in the United States. *Strix varia alleni*, the Florida barred owl, is limited to the coastal regions of South Carolina to Florida and Texas. *Strix varia albogilva*, the Texas barred owl, resides only in southern Texas. *Strix occidentalis occidentalis*, the spotted owl, and its variant, *Strix occidentalis caurina*, are the western representatives, occurring in Washington, Oregon, California, Arizona, New Mexico, and Colorado.

Owls have evoked deep-seated feelings in man throughout history. On one hand, the nocturnal habits, the chilling screams, and the humanlike stares of the owl have contributed to its reputation as a mysterious creature endowed with strange evil powers. On the other hand, the association of the owl with Athene, the Greek goddess of counsel at Athens, has contributed to its reputation as a good omen and a bird of wisdom.

Northern Pintail

Anas acuta

DESCRIPTION

Classification: Bird

Size: Length to 30″

Characteristics: *Male:* dark head, white stripe on neck, long pointed tail; *female:* mottled brown with shorter tail; both sexes white below, blue-gray feet and bill, brown speculum (wing patch) with one white border

HABITAT

Marshes

RANGE

Continental U.S.A.

IN early spring, the melting marsh ice attracts migrating waterfowl by the hundreds. One of the first arrivals, the northern pintail, seems eager to get back and settle down to breeding. Flying overhead in great flocks, wings beating rapidly and strongly, these slender, graceful birds dive sharply, level off as they near the water's surface, and drop gently into the liquid pools. This colorful display, although found throughout the United States, is most common in the western states.

The northern pintail drake is a dapper bird, his dark brown head glossed with green and purple, a white neck stripe fading into a gray back and long, thin tail. White underparts, blue-gray feet and bill, and a bronze speculum fashionably complete the attire of this genteel male. The call of the drake is a short whistle.

The female pintail duck has more subdued plumage. Streaked with dusky brown and yellow, the shorter-tailed female is smaller in size and emits a soft quack. Both males and females are strong fliers and migrate long distances.

Pintails are surface feeders, sitting high on the water with tails angled upward. Rarely diving for food, they feed on rushes, pondweed, seeds, insects, and small aquatic animals. Occasionally pintails "tip up" to reach food beneath the surface.

The courtship ritual of pintails involves social displays such as preening and body stretching. After mating, the female constructs a bowl-shaped nest on the ground among tall grass. The nest is lined with down and may be located up to one mile from the water. The female lays from seven to ten pale olive-colored eggs. The eggs are the exclusive responsibility of the female. When she leaves the nest to feed, the duck covers the clutch with the tall surrounding grass to keep them warm, moist, and hidden from view.

Soon after hatching the ducklings follow their mother to the nearest water. Brooding them alone, the female defends the ducklings courageously against predators. By July the ducklings lose their down, have complete juvenile plumage, and are on their own.

While the female is brooding the young, the adult males will molt twice. Early in the season, shortly after mating, the males develop their eclipse plumage, which resembles the female coloration. At the end of summer, the males molt again, changing from dull eclipse to bright winter plumage. By the beginning of autumn, the males initiate the southward migration, soon to be followed by the fledglings and the adult females.

The wary nature, rapid flight, and tasty flesh all contribute to the pintail's popularity as a game bird. To maintain present populations, it is protected by federal law and can be hunted only during certain seasons.

King Rail

Rallus elegans

DESCRIPTION

Classification: Bird
Size: Length to 19″; wingspan to 25″
Characteristics: Head, neck and underparts rusty; dark brown, streaked back; long, slender bill; long legs and toes; *immature:* chicks are black with white bill; young are duller in color than adults

HABITAT

Swamps, marshes

RANGE

Eastern half of the U.S.A.

DARTING easily among the medley of marsh plants, the king rail searches for food. With plumage color nearly identical to its surroundings, the secretive rail is virtually invisible and is rarely sighted unless it moves. When alarmed, the rail prefers to hide amidst the tussocks and buttonbushes rather than take wing and can remain motionless for hours. As a last resort, however, flight will ensue and the rail, with legs dangling ungainly, will fly in a low direct course over the marsh for several yards then drop abruptly into the grass. The king rail is more often heard than seen, and its call, a rapid descending *wak-wak-wak*, has a very distinctive grunt associated with it.

The king rail is a compact, hen-shaped bird with long legs. The cinnamon color of the rail is similar in both sexes, but the female is slightly smaller in size. The bill is long, slightly curved, and well adapted for consuming insects, crustaceans, worms, seeds, and berries. The body of the rail, being compressed laterally, makes it, literally, "thin as a rail." With its powerful sprinterlike legs and long toes, the rail adroitly maneuvers through the mud and submerged vegetation, preferring to run rather than fly to escape its pursuers.

The rails, nesting in groups of up to twenty pairs, are gregarious, and both sexes exhibit such courtship behavior as displaying. The nest is a deep bowl of grass, surrounded by woven grass, and is usually built on the ground or in low-lying reeds. Six to sixteen cream-colored eggs, blotched with red-brown, are laid in June. The young chicks have glossy, black down but soon develop feathers with coloration resembling the adults. Only one brood is raised each year.

King rails are quite edible, although they are not common enough to be of any economic significance. Habitat destruction has caused a decline in their population.

Common Snipe

Gallinago gallinago

DESCRIPTION

Classification: Bird
Size: Length to 11″
Characteristics: Brown with buff-striped head and back; short, orange tail; long, slender bill

HABITAT

Marshes, bogs

RANGE

Continental U.S.A.

DOZING under the thick tussock sedge, the common snipe is virtually invisible. Its brown, mottled body blending into the background, the snipe will remain motionless with barely an occasional feather stirring. As shadows lengthen, it grudgingly takes flight, uttering sharp cries as it zigzags across the marsh.

The long, slender bill of the common snipe is well suited to its feeding habits. A very sensitive instrument, the bill is used to probe mud for worms by jabbing repeatedly in a vertical direction. The snipe also relishes mosquitoes and their larvae, damselflies, mayflies, grasshoppers, crayfish, and beetles.

Migrating at night in flocks, the snipe feeds alone during the day, rejoining the flock at dusk to continue its journey. The unerring ritual to return to its former nest site each year almost spelled extinction for the species. Hunters awaiting their annual arrival killed thousands of these birds. Protected briefly by game laws, the snipe population recovered sufficiently to be returned to game bird status.

Snipes return early in the spring to mate. Winging freely in wide circles, the male snipe invites his mate to join in the exultation of the season. Together their wing and tail feathers issue a sweet hum that settles over the restless marsh. Suddenly flying low, emitting the shrill cacklings of courtship, he leads the female through the reeds to alight among the concealing grasses.

Amid the rumblings of bullfrogs and bull alligators, four olive eggs blotched and streaked in brown soon occupy a small grass-lined depression. The young snipes that break from the shells are quickly capable of foraging independently. Running over the open mud or darting through shallow water, the chicks join their mother in search of marsh delicacies.

Formerly called "Wilson's snipe" after the Scottish-American ornithologist Alexander Wilson, the common snipe is also known as meadow snipe, marsh snipe, and bog snipe.

Swamp Sparrow

Melospiza georgiana

DESCRIPTION

Classification: Bird
Size: Length to 6″; wingspan to 8″
Characteristics: Rusty crown; gray face and breast; white throat; dark eye stripe; brown back; *immature:* brown crown; streaked breast

HABITAT

Swamps, marshes, bogs

RANGE

East of the Rockies

AS the midnight sky fades to the subtler hues of gentle blues, a simple, sweet, musical trill can be heard filling the swamp. Content and secure among the sedges and reeds, the swamp sparrow clings tightly to a swaying stalk to sing its melody. One of the first to be heard in the morning and the last to fall silent at night, the sparrow seems equally eager to welcome the day as it does to bid it farewell.

Timid and retiring, the stocky swamp sparrow rarely ventures far from the wetlands. When flushed from its perch, it flutters hastily into a clump of thick reeds, where it remains determinedly lodged and resolutely silent. By producing squeaking noises, a patient observer may be able to lure the sparrow out to sing once again.

The swamp sparrow is somewhat smaller than its relative, the song sparrow. With gray face and breast, white throat, rusty crown, dark eye stripes, and a brown back, the swamp sparrow blends well with its surroundings. The female sparrow is smaller than the male and is similar in color, while the young have streaked breasts and brown crowns. The adult plumage is usually assumed by the first fall molt.

Feeding mainly on aquatic insects and weed and grass seeds, the swamp sparrow forages on the ground. During the warmer months it inhabits the wetlands from the central eastern United States north. It may be found south to Florida and the Gulf coast in upland areas during the colder months.

After mating, the female lays four to five blue-green, brown-spotted eggs in a cup-shaped nest made of grass. The nest is usually located on the ground or in a mass of tangled sedge. The female incubates the eggs from twelve to fifteen days and may raise from one to two annual broods. The chicks may remain in the nest for up to thirteen days until fully fledged. Occasionally a nestling attempting its maiden flight plunges into the water below and meets an untimely death in the form of a hungry fish.

The swamp sparrow is of no great economic importance because it does not feed near commonly cultivated areas. Since its food is mainly weed seeds and insects, however, it is certainly important in the natural food web to keep such organisms in check.

Marsh Wren

Cistothorus palustris

DESCRIPTION

Classification: Bird
Size: Length to 5½″; wingspan to 7″
Characteristics: Brown above with black crown; white line over eye; black back with white stripes

HABITAT

Marshes

RANGE

East of the Rockies

CREEPING down a cattail stem to snatch a juicy mosquito larva from the water, the brown marsh wren strikes quickly and precisely. This shy, secretive bird is fond of deep marshes and flits restlessly from branch to branch, alert yet apprehensive about its surroundings. Cocking its tail when excited, the bird disappears easily into the tall, green reeds it calls home and will scold an intruder with a harsh, constant note. The marsh wren has a penchant for song and will spend many days and nights filling the marsh with its sweet, liquid voice.

The marsh wren is a small, brown bird with a black and white streaked back, a black crown, and a turned-up tail. It has a white stripe over its eye, and its underparts vary from cream to cinnamon brown. Males and females are similar in coloration.

There are many varieties of marsh wrens distributed over the United States, one of the most closely related being *Cistothorus platensis*. This variety is found mainly in bogs and wet meadows.

As spring debuts, so does the male marsh wren, who arrives first to erect dummy nests. When the female arrives several weeks later, the male, who is often polygamous, displays for a mate or mates. Fluffing out his breast feathers proudly and haughtily cocking his tail, the marsh wren sings loudly. Upon acceptance by a female, the male will mate with her. Immediately following this encounter, the female proceeds to make her own nest, ignoring the efforts of the male. About the size of a baseball, the nest is constructed of woven plants tightly bound to cattail stems. With a side entrance and inner doorstep, it is lined with feathers and cattail down. From five to nine red-brown speckled, chocolate-colored eggs are laid and incubated for thirteen days. Up to two broods may be raised in a single season.

The habit of excessive nest construction, characteristic of many wrens, has provoked a number of theories. One is that the construction of a large quantity of nests may confuse snakes and other predators. Another theory is that male wrens need to work off their inherent nervousness by building many nests. No conjecture has, of yet, adequately explained this strange behavior.

The food of the marsh wren consists mainly of insects found among the marsh plants. In this way the bird helps to keep the insect population of the wetlands under control. The diminutive marsh wren is preyed upon by many animals, and its nest is often overtaken by bees and tree frogs. Still the cheerful marsh wren perseveres to brighten the wetlands with its pleasing melody.

Beaver

Castor canadensis

DESCRIPTION

Classification: Mammal
Size: Length to 4′; weight to 70 lbs.
Characteristics: Stout body; dark brown fur; flat paddle-shaped tail

HABITAT

Marshes

RANGE

Continental U.S.A.

WHILE few of us have ever seen a beaver in the wild, the signs of its presence are unmistakable. A dam of sticks and mud flooding a stream, gnawed stumps, and felled trees all suggest the activities of this elusive rodent.

Beaver pelts were so valued in the seventeenth and eighteenth centuries that the European species were nearly hunted to extinction. England, having no natural populations of its own, exploited the market by sending trappers to the New World. The famed Hudson's Bay Company flourished largely on the beaver pelt trade, and soon the North American beaver was threatened with extinction. In the early nineteenth century, conservation groups, recognizing a broader value in this animal, influenced legislators to initiate protective measures. Since that time, both local and national legislation in the United States, Canada, and northern Europe has effected a successful increase in beaver populations.

A beaver lodge is of such high engineering quality that beavers have been credited with far more native intelligence than they are due. The lodge is made of carefully laid sticks packed with mud and stones. While the lodge is a visible dome above the water's surface, at least half of the lodge is beneath the water level. Within, there is a conical central chamber above the water with at least two different points of entry or escape under water.

Beavers are among the largest North American rodents. Their dense undercoat and thick outer fur are preened with a natural body oil that protects the animals from the icy winter waters that they inhabit. Valves in their nostrils and ears are closed securely when the animals enter the water. Aided by webbed back feet and a large flat tail, they are quite proficient swimmers.

Aspen, willow, poplar, and birch trees are the main sources of nutrition for beavers. They eat the bark and sapwood of the twigs and smaller branches, a diet quite rich in carbohydrates and proteins. The large incisor teeth are constantly worn down as the beaver gnaws and fells trees. Throughout its life, however, these teeth experience continuous growth, thereby maintaining a rather uniform length.

Beavers are monogamous mammals. The young are born anytime during the spring and remain as a family for two years. During those years much information seems to be passed from parents to young, especially that pertaining to the repair and maintenance of the lodge.

Southern Bog Lemming

Synaptomys cooperi

DESCRIPTION

Classification: Mammal

Size: Length to 4″

Characteristics: Brown fur; shorter tail and legs, smaller eyes and ears, and stouter body than mice

HABITAT

Swamps, marshes, bogs

RANGE

Throughout the Northeast except certain sections of Massachusetts, Rhode Island, Connecticut, New York, and New Jersey

TUCKED safely in a globular nest of dried grasses, three newborn bog lemmings are gently nuzzled awake by their mother and begin to nurse vigorously. It matters not that their tightly clamped eyes preclude visions of this moment, for their nest is six inches underground, blocking even the most penetrating rays of sunshine. The nest, carefully dug near an obscure system of tunnels, has several entrances from within and below the peatland vegetation. Here the litter will be safe from the pursuits of predatory mammals and owls, although the site can offer no protection from a determined snake.

The chubby, stubby bog lemming, active day and night throughout the year, burrowing through vegetation and navigating underground burrows, is rarely seen by the wetlands naturalist. Lacking the charisma and Disney appeal of his many rodent cousins,he scurries through life on abbreviated legs, clipping grasses into one-inch lengths and tidily stacking them along his routes.

These miniature haystacks and prominent bright green droppings reveal the presence of these sedulous creatures. Bog lemmings, also called bog voles, consume great quantities of grasses, especially bluegrass and clover, to maintain their prodigious caloric output. Seeds and the stems of herbaceous plants supplement their dietary needs.

The unusual fluctuations of the yearly populations have prompted much inquiry. It has been suggested that, while there is no evidence of the reputed suicide trips to the sea, the stress of an extremely high metabolism accounts for the bog lemming's very short life span. They may literally wear themselves out and then die.

Whatever the reason, the reticent bog lemming passes but briefly through this earthly life in a flurry of activity, leaving in his wake tiny stacks of hay and even tinier clouds of dust.

Mink
Mustela vison

DESCRIPTION

Classification: Mammal
Size: length to 25″
Characteristics: Dark brown fur; white chin patch; pointed nose; small ears; bushy tail

HABITAT

Marshes

RANGE

Continental U.S.A. except Arizona and parts of Utah, California, New Mexico, and Texas

EARLY in his evolutionary history, man learned the fine art of animal husbandry. Of course, the better part of this practice centered upon dietary considerations and the exploitation of brawn. Domestication, however, soon revealed other uses. Since the nineteenth century, mink, among the most highly prized furbearers, have been raised commercially for their rich, brown pelts. Today nearly 40 percent of the mink fur crop is harvested from captive breeding farms. Scientifically designed selective breeding programs have produced the broad spectrum of soft, lustrous pelt colors now available to prospective customers.

In the wild, mink prefer the solitude of cattail marshes. Spending much of their time in the water, they are generally safe from their primary predators, foxes, bobcats, and great horned owls. Their slightly webbed hind toes enhance their excellent swimming skills.

The mink's naturally inquisitive nature often leads to violent confrontations. Males are particularly fierce fighters, and both sexes scream, hiss, and spit when threatened. Males emit a strong odor when cornered, and although they cannot spray, the fetid odor is much stronger than the familiar release of the skunk.

Mink dens are usually temporary lodgings. They are often abandoned muskrat burrows or hollow logs. Mink also dig dens between tree roots along water margins. During the early spring a den may be inhabited by a female and her newborn litter. Since both sexes accept more than one mate the presence of a male in the family unit is inconsistent. Mink produce only a single litter each spring, and while the young can survive alone by early summer, they will often remain with their mother into the fall.

The carnivorous mink kills its prey by sinking sharply pointed teeth into its victim's neck. Frequently killing more than a single appetite can consume, the mink then stores the remains in its den. Muskrats, mice, fish, and marsh birds provide the bulk of the mink's nutritional needs.

Star-Nosed Mole

Condylura cristata

DESCRIPTION

Classification: Mammal
Size: Length to 8″ (including tail)
Characteristics: A ring of 22 fleshy, pink tentacles at end of nose; dark brown or black fur

HABITAT

Freshwater margins

RANGE

Maine to Georgia; west to Illinois and Ohio

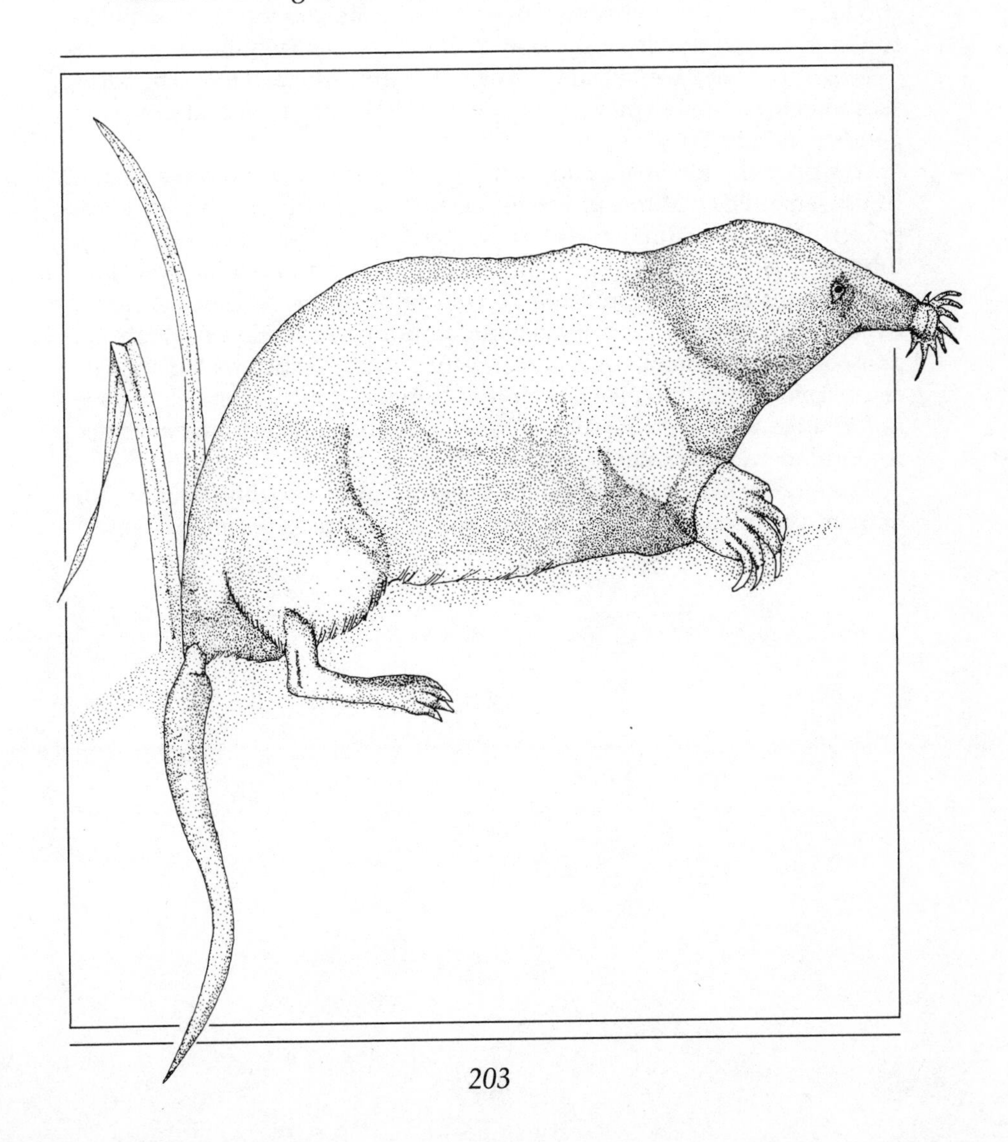

THE star-nosed mole derives its name from the series of fleshy tentacles arranged in a star or disklike pattern around its mouth. Unhampered by its limited vision and poor sense of smell, the mole's keen hearing and the tactile sensitivity of the snout help it to maintain a varied diet. It usually feeds both day and night on earthworms, aquatic insects and larvae, small fish, crustaceans, and mollusks. Its high metabolic rate is life-threatening if the animal fasts for more than half a day. Moles, known for their larders, store supplies of earthworms in subterranean burrows. During the winter their tails appear enlarged from deposits of stored fats.

Pinpoints of eyes peak through the velvety soft fur, which fully conceals the mole's ears. This fur is arranged so that it is apparently undisturbed by the animal's backward movement underground. The stubby, muscular body is easily elongated to negotiate narrow tunnels. The shovellike forefeet and heavy claws allow the mole to move through the soil in the same breast-stroke fashion it uses while swimming.

Moles pair in late fall and remain together during their breeding season. Three to six young are commonly born in April and are independent within three weeks.

It is only the fortunate naturalist who happens upon a mole. These wary, sensitive creatures prefer to remain underground. The small but powerful shoulders may direct the excavation of three hundred feet of tunnels in a single day. The conspicuous ridges, formed by tunneling near the surface, are more likely to be seen than the mole or its tracks. Occasionally a mole may surface and continue tunneling through tall grass or snow. Molehills are characteristic among the burrowing animals as the pile of earth expelled from the tunnel. The tunnels of the star-nosed mole, however, rarely have such openings. Instead the entrance is more likely to be found underwater.

The star-nosed mole's economic insignificance probably accounts for its relative obscurity. Pursuing its prosaic activities, the mole is most likely content with this oversight.

Muskrat
Ondatra zibethicus

DESCRIPTION

Classification: Mammal
Size: Length to 24″ (including tail)
Characteristics: Chocolate brown fur above, grayish below; long, sparsely haired tail; short round ears

HABITAT

Swamps, marshes

RANGE

Continental U.S.A.

MUCH like the beaver and otter, the muskrat is well suited for an aquatic lifestyle. Adaptations such as partially webbed hind feet, folds of skin protecting inner ears, and a laterally flattened rudderlike tail all enhance its excellent performance in the water.

A marsh or densely vegetated swamp may be the site of great mounds of nonwoody plant stems carefully interlaced above the water level. Muskrat houses may be as large as four feet by six feet with a nest snuggled safely in the center. Around the home may be other signs of this aquatic mammal. Tunnels cleared through vegetation, evidence of slides along steep banks, and distinctive tail tracks marked in the mud are all telltale signs.

Muskrats consume great volumes of aquatic vegetation, especially wild rice, cattails, and arrowheads. Although they appear to store food at times, they have been observed digging in the winter mud for the roots of their normal fair-weather fare. Muskrats supplement their diets rather regularly with freshwater mussels and crayfish.

The pelts of North American muskrats are coveted by trappers. The rich brown fur experiences no seasonal changes. It is retailed as Hudson seal, Russian otter, red seal, and river mink. Muskrats are also sought for their reportedly tasty meat, which is commercially marketed as marsh rabbit and Chesapeake terrapin. Although it is a prolific breeder, the muskrat suffers reduced numbers where hunting and trapping are unchecked.

In the north, breeding occurs from March through September. Three to four litters per season are not uncommon. In the southern climates, however, breeding continues throughout the year. Four to nine young are born after a twenty-nine-day gestation period. The young rats are quite helpless, born furless and with eyes closed. Within a month they will be weaned, evicted from the nest, and ready to assume the life struggles common to adulthood.

One of the greatest dangers that these animals face is flooding. Normally they prefer to inhabit stable waterways. But when driven from their homes by floods they become quite vulnerable to predation, disease, cold, and intense competition for food. These are courageous animals whose small size does not deter them from facing up to foxes, raccoons, hawks, owls, or even eagles.

Their aquatic constructions may lack the intricacies of a beaver house, their personalities may fall short of the raccoon's appeal, and their pelts may lack the public relations of mink and otter, but muskrats offer a steadfast durability—a teeth-gritting determination—that is one of our wetlands treasures.

Raccoon

Procyon lotor

DESCRIPTION

Classification: Mammal
Size: Length to 34″; tail to 10″; weight to 35 lbs.
Characteristics: Fur gray to brown; black band across face and eyes; 6–7 black rings on bushy tail

HABITAT

Swamps

RANGE

Continental U.S.A.

HIGH in the hollow of an aged bald cypress, a raccoon grasps the bark with deft little hands and creeps backward down the tree. Three smaller versions follow their mother cautiously, while she remains vigilantly alert for danger. Once the family has reassembled on the ground, the female will lead her young on their nightly foraging expedition. Raccoons enjoy a varied diet, making excellent use of the foods available with each season. Gleaning the wetlands of insects, crayfish, eggs, and vegetation, the troop ambles along with heads bent toward the ground, backs arched, and tails dragging.

Raccoons are basically sedentary animals. A sharp-eyed naturalist may spy a raccoon sunning on the branch of a mature tree. But more likely only the telltale signs of raccoon activity, scratches along the bark of a hollow tree or crayfish remains along a water margin, will be noticed.

The very social and inquisitive nature displayed by young raccoons has made them the subject of exaggerated tales of domestication. While the young are easily tamed, they often revert to feral behavior when they sexually mature at one year. Raccoons are excellent climbers and fighters and will attack viciously if they anticipate danger. Males, especially, are somewhat territorial and will usually greet another male raccoon with bristling back hairs, posturing, and growling.

Each winter, as early as December in the south, mature females seek out hollow trees, culverts, or rock clefts to build their nests. While females accept only one male per season, a single male may breed several females. Two to six young are born in each litter between April and May. The family unit, which may include the male, is maintained through the following winter if the young can avoid the predation of foxes, bobcats, and owls.

During the fall, foraging activities increase as the raccoons prepare their fat reserves for the demands of winter. Acorns, high in calories, are especially desirable.Throughout the winter, an animal may lose up to 50 percent of its fall weight. Raccoons do not actually hibernate, but they may sleep several days at a time to withstand the most bitter temperatures.

The raccoon's most notable behavior, washing its food before eating, has inspired both its common and scientific names. *Lotor* is of Latin origin meaning "washes." The Algonquin Indians described the masked creature as "aroughcoune," that is, "he who scratches with his hands." Careful observation of these animals suggests that the water serves to enhance the raccoon's tactile sense. After dipping its food in water, the raccoon usually scratches, kneads, and generally manipulates the food item for several minutes.

Since the nineteenth century, raccoon pelts have been highly regarded. The long, high-quality fur is marketed as "Alaska bear," "Alaska sable," or simply as raccoon. Attempts to establish commercial raccoon farms in Canada and Europe have failed mainly due to the availability and low cost of trapping raccoons in America.

Appendix

Wetland Areas in the National Park and National Wildlife Refuge System

1. Moosehorn National Wildlife Refuge (Maine)

Moosehorn, established in 1937, has 22,565 acres. It is made of habitats ranging from upland areas to lakes, streams, marshes and tidal flats. On the Cobscook Bay side, where tides may reach twenty-eight feet, are harbor seals and porpoises. The remainder of the refuge contains predominantly spruce, larch and cedar trees, and abounds in songbirds and mammals. Containing some sixty miles of trails, including several with self-guiding leaflets, this area is a favorite stopover of migrating ducks and geese.

Box X, Calais, Maine 04619

2. Acadia National Park (Maine)

Wetland areas in this 41,642-acre park, including Great Meadow, Gilmore Meadow, and Newmill Meadow, are easily accessible by a well kept system of foot trails. Big Heath, on Mt. Desert Island, is one of the most striking examples of a black spruce forest in the National Park System. Nestled in the center of the encircling towering spruces, this bog sports an extensive sphagnum moss cover dotted liberally with carnivorous pitcher plants. Other bogs, in varying stages of development, can also be found scattered throughout the park.

Route 1, Box 1, Bar Harbor, Maine 04609

3. Cape Cod National Seashore (Massachusetts)

Containing both fresh and saltwater marshes, as well as bogs, this area attracts thousands of visitors each year. Red Maple Swamp hosts an abundance of bird life and dense stands of sweet pepper bush and green briar. Swamp azalea and highbush blueberry thrive in Eastham Bog. This area also contains the Atlantic White Cedar Swamp and the Herring River Marsh. On Nauset, with its large tidal flats and lush grass, will be found algae, worms, sponges, crabs, and clams.

South Wellfleet, Massachusetts 02663

4. Great Swamp (New Jersey)

This 6000-acre area is almost within view of the Manhattan skyline. Though surrounded by development, it is possible to get lost in this land of bogs, wooded swamps, freshwater marshes, and uplands. White-tailed deer abound in this area as do wood ducks. The refuge has more than two hundred species of birds and plays host to warbler and raptor migrations. It also contains the rare blue-spotted salamander and bog turtle along with many diverse botanical species. Auto drives and hiking trails are available for visitors.

Basking Ridge, New Jersey 07920

5. Blackwater National Wildlife Refuge (Maryland)

Blackwater has 11,800 acres and is composed of almost 80 percent marsh and water with the rest being mixed hardwood and pine woods. It has an observation tower, woodland walking trails, and a four-mile auto tour route from which most of the refuge wildlife can be seen. Ducks and geese are one of the prime sights, and at the peak of fall migration more than 150,000 waterfowl stop here. Three species of southern bald eagle nest at this refuge as well as many marsh birds.

Route 1, Box 121, Cambridge, Maryland 21613

6. National Capital Parks (District of Columbia, Virginia, Maryland)

In a 7,054-acre area around the nation's capital there are several interesting wetlands. One is called Dyke Marsh and is an extensive cattail marsh along the Mount Vernon Parkway near Belle Haven. On Theodore Roosevelt Island in the Potomac River is a cattail marsh and a silver maple swamp. Open to the public in the summer months are several swamps along the Anacostia River, each with guided nature walks.

1100 Ohio Drive, SW, Washington, D.C. 20242

7. Cape Hatteras National Seashore (North Carolina)

Fresh and saltwater marshes abound on this seventy-mile-long chain of barrier islands. Many snow geese, ducks and whistling swans spend their winters in the freshwater marshes on the north end of Hatteras Island. Tidal salt marshes, containing smooth cordgrass, salt grass, black rush, and marsh alder, are found along the west side of the island.

P.O. Box 457, Manteo, North Carolina 27954

8. Okefenokee National Wildlife Refuge (Georgia)

Occupying about 80 percent of the Okefenokee Swamp, this refuge is one of the oldest and most primitive areas in the United States. The name is the white man's version of the Indian words for "Land of the Trembling Earth." Containing thick and unstable peat floors, the cypress swamps are known for their quaking, as visitors walk upon them. Bears, raccoons, and muskrats live in the swamp as well as many species of water birds, such as the ibis, wood duck, and anhinga. One of the largest alligator populations anywhere, some twelve to fifteen thousand, is found here. Carnivorous bladderworts and pitcher plants, marsh marigolds, water lilies, and pickerelweed can also be found here.

P.O. Box 117, Waycross, Georgia 31501

9. Everglades National Park (Florida)

Teeming with wildlife, the Florida everglades are the largest American subtropical wilderness. A transition zone between temperate and tropical climates, this area abounds in exciting ecological phenomena. Visitors can take advantage of guided swamp hikes, boardwalks through the mangrove, and tram rides in search of alligators and aquatic birds. The Everglades are a winter rest area for teal, pintails, and wood ducks.

P.O. Box 279, Homestead, Florida 33030

10. Reelfoot National Wildlife Refuge (Tennessee, Kentucky)

Reelfoot Lake is the seasonal home of extensive numbers of bald eagles, waterfowl, and migrating songbirds. Containing virgin bald cypress stands, this 10,142-acre refuge is one of the most picturesque areas in the nation. Mammals are readily seen, as are reptiles and amphibians. Visitors can access both hiking and auto trails.

Box 98, Samburg, Tennessee 38254

11. White River National Wildlife Refuge (Arkansas)

A tributary of the Mississippi River, this refuge is a haven for king rails, marsh wrens, and herons. During the spring rains small boats are available for touring the area, while hiking along footpaths is customary the rest of the year. Nesting wood ducks, black bears, mink, and beavers are abundant.

P.O. Box 308, De Witt, Arkansas 72042

12. Aransas National Wildlife Refuge (Texas)

Composed of 90,000 acres, Aransas is one of the few places where whooping cranes can be seen. A recent count showed almost one hundred cranes with a few more in small captive groups. This is a dramatic change since 1923 when the whooping crane was almost extinct. Aransas has well kept trails from which to view a great variety of flora and fauna, including over 350 species of birds, white-tailed deer, alligators, coyotes, and mountain lions.

P.O. Box 100, Austwell, Texas 77950

13. Upper Mississippi River Wildlife Refuge (Minnesota)

The best way to see this 284-mile-long refuge is by boat, although there is a continuous system of highways closely following its boundaries. Some of the wildlife found in the area include deer, wood ducks, barred owls, and red-tailed hawks.

122 West Second Street, Winona, Minnesota 55897

14. Seney National Wildlife Refuge (Michigan)

Seney, a 153-square-mile refuge, was built on reclaimed land. It is ideal for wildlife with over seven thousand acres of open water interlaced with fields, islands, and forested ridges. Guided auto trails and miles of foot trails allow visitors the opportunity to view over two hundred species of birds and small mammals. Fishing and hunting are permitted with proper authorization.

Star Route, Seney, Michigan 49883

15. Isle Royale National Park (Michigan)

On an island in Lake Superior the 210-square-mile park includes two types of bogs, an extensive acid black spruce swamp, and a more alkaline-based white cedar swamp. Visitors may observe bogs in various stages of development, sedge mats, sphagnum moss, white cedar, and wandering resident moose.

87 North Ripley Street, Houghton, Michigan 40031

16. Lower Souris and Upper Souris National Wildlife refuge (N. Dakota)

These two refuges are located on the Souris River and extend along the valley. The Upper Souris is the nesting site of whistling swans, western grebes, and many species of ducks. Deer, rodents, and raptors are also common to this area. The Lower Souris is an important stopover for the central flyway. Raccoons, beavers, minks, and muskrats abound here. Visitors have access by both auto and foot trails.

RR 1, Foxholm, North Dakota 58738

17. Glacier National Park (Montana)

The 1,500 square-mile-park is similar to central Alaska in climate. Marshes and bogs have developed in this portion of the Rocky Mountains between the creeks, rivers, and lakes. Guided walks are available for visitors.

West Glacier, Montana 59936

18. Red Rock Lakes National Wildlife Refuge (Montana)

Red Rock Lakes contains sixty-two square miles, much of which is covered by shallow lakes, marshes, and meadows. It is here that the trumpeter swan now survives and is protected. The fields and mountains encourage and support an abundant diversity of wildlife. Although the refuge is accessible in summer by auto and foot trails, it is closed in winter.

Monida Star Route, Box 15, Lima, Montana 59739

19. Yellowstone National Park (Idaho, Montana, Wyoming)

The marshes in Yellowstone are inhabited by a variety of wildlife. Beavers and moose are frequently sighted. The marshes may be visited by foot trails, auto, or horseback. Featuring the world's most spectacular displays of geothermal water, parts of Yellowstone are open to the public year round.

Wyoming, 82190

20. Grand Teton National Park (Wyoming)

Six miles south of Yellowstone, visitors are treated to a wide variety of wetlands flora and fauna. Ospreys and trumpeter swans abound in the marshes and waterways. The foot paths are edged with delicate wildflowers and shrubs. Those who prefer not to walk may explore by boat, horseback, or skis.

P.O. Box 67, Moose, Wyoming 83012

21. Yosemite National Park (California)

Perched in the Sierra Nevada Mountains, this park is largely wilderness. Carved by glaciers, many meadow marshes have developed along the streams and lakes. Visitors are treated to stands of giant sequoias and nearly three hundred species of birds and mammals.

P.O. Box 577, National Park, California 95839

22. Tule Lake, Lower Klamath, and Sacramento National Wildlife Refuges (California)

Two hundred fifty species of waterfowl, waders such as the sandhill crane, and hundreds of bald eagles are regular visitors and residents of these refuges. From tule marshes to coniferous forests, the varied ecological niches make this area a unique wildlife preserve. Camping facilities and canoe trails make the Upper Klamath quite accessible to the amateur naturalist.

Route 1, Box 74, Tule Lake, California 96134

23. Malheur National Wildlife Refuge (Oregon)

Encompassing 282 square miles, Malheur is a sampling of marshes, ponds, riverbottoms, lakes, and desert. Over 260 species of birds and 57 species of mammals reside here. Malheur is noted for its immense variety of wildlife including eagles, herons, terns, swans, hawks, mule deer, and coyotes. In the fall thousands of waterfowl migrate through this refuge, which is managed with precisely that goal. There are many auto trails, foot paths, and canoe routes for visitors to access much of the area. The park is closed through the winter.

P.O. Box 113, Burns, Oregon 97720

Glossary

achene: a small, dry, one-seeded fruit
annual: a plant that completes its life cycle from seed germination to production of a new seed in one growing season
anther: part of the stamen in a flowering plant that when ripe splits to release pollen
barbel: a slender tactile extension near the mouth of certain fish
bract: a small leaflike structure
buttress root: a type of root that grows near the main plant stem of certain trees
carapace: the upper shell of a turtle
catkin: a spike of unisexual flowers that often hang downward
chelonian: resembling or relating to turtles and tortoises
cleistogamous: exhibiting inconspicuous nonopening self-pollinating flowers
conifer: a group of trees or shrubs that bear woody cones
corm: an underground, swollen, rounded base of a plant stem that stores food
crozier: the young, coiled frond of a fern
dimorphism: exhibiting two different forms
emergent: a plant rooted in water and having most of the vegetative growth above water
fiddlehead: the young unfurling fronds of certain ferns
filter feeding: a method of feeding found in many aquatic invertebrates in which minute organisms are filtered from the surrounding water by specialized tissue
frond: the leaflike structure of a fern
glochidial stage: the larval form of certain freshwater mussels, which is discharged into the water and attaches itself as an external parasite to the gills or fins of certain fish
harrier: any of various slender hawks with long, angled wings feeding chiefly on small mammals and reptiles, which hunt by flying low over open ground
hemotoxic: causing dissolution of the red blood cells
hermaphrodite: a flowering plant or animal having both male and female structures on the same individual
hummock: a rounded or conical knoll of vegetation
indusium: an outgrowth of a leaf which covers the sori in many ferns
lepidoptera: the family of insects including the moths and butterflies
nutlet: a small nutlike fruit
once cut: leaf or frond cut into a number of simple leaflets
panicle: a pyramidal, loosely branched flower cluster
perennial: a plant that takes many years to reach full size and begin to reproduce
pinna: a segment of a compound leaf

pinnate: a compound leaf characterized by leaflets arranged on opposite sides of an axis
pinnule: a secondary pinna or a division of a compound leaf
pistil: the collective name for the female reproductive organs in a flower
plastron: the lower shell of a turtle
proboscis: a long, tubelike structure that projects from the head of an animal
prothallus: a small, flat, green vegetative body attached to the soil by filamentous strands and arising from the spores of certain lower plants
raceme: elongated flower cluster with stalked flowers along a central stem
raptor: a bird of prey
rhizome: a horizontal underground plant stem that bears buds, leaves, and roots; a means of vegetative reproduction
rookery: the breeding colony of certain birds
samara: dry, one-seeded winged fruit
sepals: leaflike structures, usually green, which surround the colored petals of a flower
sori: clusters of sporangia on a fertile frond of a fern
spadix: a type of flower cluster enclosed and protected by a large bract that surrounds it
speculum: a splash of color on certain quill feathers of ducks and certain other waterfowl
spermatophore: a capsule enclosing spermatozoa extruded by the male of various animals and functioning in the insemination of the female
spermatozoa: sperm
sporangium: a structure that produces asexual spores
stamen: the male reproductive organ of a flower
stigma: the terminal portion of a flower's female reproductive organ to which the pollen grains adhere during pollination
thallus: a plant body not clearly differentiated into stem and leaf, often without roots or rhizoids
thoracic: referring to the thorax; the body segment of an insect between the head and abdomen
transpiration: the loss of water from the surface of land plants
twice cut: a leaf or frond cut into a number of simple leaflets, which in turn are cut into subleaflets
tympanum: vibratory membrane covering the external opening of the middle ear of most amphibians
umbel: a type of flower cluster in which the flower stalks arise from the same point

Bibliography

Angier, Bradford. *Field Guide to Medicinal Wild Plants.* New York: Stackpole Books, 1978.

Audubon Nature Encyclopedia. (National Audobon Society.) New York: Curtis Books, 1973.

Blaustein, Elliot. *Name That Fern.* Franklin Lakes: Saffyre Publications, 1979.

Brockman, Frank C. *Trees of North America.* New York: Golden Press, 1968.

Brown, Lauren. *Grasses: An Identification Guide.* Boston: Houghton Mifflin, 1979.

Brown, Vinson. *Knowing the Outdoors in the Dark.* New York: Collier-Macmillan, 1972.

Bull, John, and Farrand, John. *The Audubon Society Field Guide to North American Birds.* New York: Alfred A. Knopf, 1977.

Burton, Maurice, and Burton, Robert, eds. *International Wildlife Encyclopedia.* New York: Marshall Cavendish Corp., 1969.

Carr, Archie. *Handbook of Turtles.* Ithaca: Cornell University Press, 1978.

Clement, Roland C. *Birds.* New York: Bantam Books, 1973.

———. *Nature Atlas of America.* Maplewood: Ridge Press and Hammond, 1973.

Coates, Alice M. *Flowers and Their Histories.* New York: McGraw-Hill Book Company, 1968.

Cobb, Boughton. *A Field Guide to Ferns.* Boston: Houghton Mifflin, 1963.

Collins, Henry Hill. *Complete Field Guide to North American Wildlife.* New York: Harper and Row, 1981.

Collins, Stephen. *The Biotic Communities of Greenbrook Sanctuary.* New Brunswick: Stephen Collins, 1957.

Dana, William Starr. *How to Know the Wild Flowers.* New York: Dover Publications, 1963.

Feibleman, Peter S. *The Bayous.* Alexandria: Time-Life Books, 1977.

Foster, Gordon F. *Ferns to Know and Grow.* New York: Hawthorne Books, 1975.

Gabriel, Ingrid. *Herb Identifier and Handbook.* New York: Sterling, 1975.

Gibbons, Euell. *Stalking the Healthful Herbs.* New York: David McKay Company, 1966.

Harding, John J., ed. *Marsh, Meadow, Mountain—Natural Places of the Delaware Valley.* Philadelphia: Running Press, 1977.

Harlow, William M. *Trees of the Eastern and Central United States and Canada.* New York: Dover Publications, 1957.

Headstrom, Richard. *Suburban Wildflowers.* Englewood Cliffs: Prentice-Hall, 1984.

———. *Suburban Wildlife.* Englewood Cliffs: Prentice-Hall, 1984.

Hinds, Harold R., and Hathaway, Wilfred A. *Wildflowers of Cape Cod.* Massachusetts: Chatham Press, 1968.

Kiernan, John. *An Introduction to Wild Flowers.* New York: Doubleday and Company, 1952.
Knobel, Edward. *Field Guide to Grasses, Sedges and Rushes.* New York: Dover Publications, 1980.
Lawrence, Gale. *A Field Guide to the Familiar.* Englewood Cliffs: Prentice-Hall, 1984.
Little, Elbert L. *Forest Trees of the United States and Canada, and How to Identify Them.* New York: Dover Publications, 1979.
Milne, Louis J., and Milne, Margery. *The Mystery of the Bog Forest.* New York: Dodd, Mead and Company, 1984.
Mitchell, John, and the Massachusetts Audubon Society. *The Curious Naturalist.* Englewood Cliffs: Prentice-Hall, 1980.
Montgomery, F. H. *Trees of the Northern United States and Canada.* New York: Ryerson Press, 1970.
Muenscher, Walter C. *Poisonous Plants of the United States.* New York: Macmillan Publishing Company, 1975.
Newcomb, Laurence. *Newcomb's Wildflower Guide.* Boston: Little, Brown and Company, 1977.
Nicholls, Richard E. *The Running Press Book of Turtles.* Philadelphia: Running Press, 1977.
Niering, William A. *The Life of the Marsh.* New York: McGraw-Hill Book Company, 1966.
———, and Goodwin, Richard. *Inland Water Plants of Connecticut.* New London: Connecticut College Press, 1973.
Palmer, Laurence. *Fieldbook of Natural History.* New York: McGraw-Hill Book Company, 1949.
Parsons, Frances Theodora. *How to Know the Ferns.* New York: Dover Publications, 1961.
Pearson, Gilbert T. *Birds of America.* New York: Doubleday and Company, 1936.
Perrins, Christopher, and Harrison, C. J. O., eds. *Birds: Their Life, Their Ways, Their World.* New York: Reader's Digest Association, 1979.
Peterson, Lee Allen. *A Field Guide to Edible Wild Plants.* Boston: Houghton Mifflin, 1979.
Peterson, Roger Tory. *A Field Guide to the Birds.* Boston: Houghton Mifflin, 1980.
———, and McKenny, Margaret. *A Field Guide to the Wildflowers.* Boston: Houghton Mifflin, 1968.
Petrides, George. *A Field Guide to Trees and Shrubs.* Boston: Houghton Mifflin, 1972.
Petry, Loren C., and Norman, Marcia G. *A Beachcomber's Botany.* Massachusetts: Chatham Conservation Foundation, 1968.
Pond, Barbara. *A Sample of Wayside Herbs.* Connecticut: Chatham Press, 1974.
Prince, J. H. *Plants That Eat Animals.* New York: Thomas Nelson, 1979.
Pringle, Laurence. *Wild Foods.* New York: Four Winds Press, 1978.
Robbins, Chandler S.; Bruun, Bertel; and Zim, Herbert S. *Birds of North America.* New York: Golden Press, 1966.
Samson, John G. *The Pond.* New York: Alfred A. Knopf, 1979.
Schaeffer, Elizabeth. *Dandelions, Pokeweed and Goosefoot.* New York: Addison-Wesley, 1972.
Sears, Paul B. *Lands Beyond the Forest.* Franklin Lakes: Saffyre Publications, 1979.
Shelford, Victor E. *The Ecology of North America.* Chicago: University of Illinois Press, 1972.

Shuttleworth, Floyd S., and Zim, Herbert S. *Non-Flowering Plants.* (Golden Nature Guide.) New York: Golden Press, 1967.

Simon, Carol. *Reptiles.* New York: Dodd, Mead and Company, 1988.

Smith, Hobart M., and Barlowe, Sly. *Amphibians of North America.* (Golden Nature Guide.) New York: Golden Press, 1978.

Smith, Robert Leo. *Ecology and Field Biology.* 2d ed. New York: Harper and Row, 1965.

Socha, Laura. *A Bird Watcher's Handbook.* New York: Dodd, Mead and Company, 1987.

Stephens, H. A. *Poisonous Plants of the Central United States.* Kansas City: Regents Press of Kansas, 1980.

Stokes, Donald W. *A Guide to Nature in Winter.* Boston: Little, Brown and Company, 1976.

———. *The Natural History of Wild Shrubs and Vines.*New York: Harper and Row, 1981.

Stupka, Arthur, and Robinson, Donald. *Wildflowers in Color.* New York: Harper and Row, 1965.

Walker, Laurence. *Trees.* Englewood Cliffs: Prentice-Hall, 1984.

Weiner, Michael A. *Earth Medicine—Earth Food.* New York: Macmillan, 1980.

Weller, Milton. *Freshwater Marshes.* Minneapolis: University of Minnesota Press, 1981.

Wernert, Susan J., ed. *North American Wildlife.* New York: Reader's Digest Association, 1982.

Whitney, Stephen. *Western Fronts.* New York: Alfred A. Knopf, 1985.

Wilson, Ron, and Lee, Pat. *The Marshland World.* Dorset: Blandford Press, 1982.

Zim, Herbert S., and Martin, Alexander C. *Flowers.* (Golden Nature Guide.) New York: Golden Press, 1950.

Index